Distillation Processes - From Solar and Membrane Distillation to Reactive Distillation Modelling, Simulation and Optimization

Edited by Vilmar Steffen

Published in London, United Kingdom

IntechOpen

Supporting open minds since 2005

Distillation Processes - From Solar and Membrane Distillation to Reactive Distillation Modelling, Simulation and Optimization
http://dx.doi.org/10.5772/intechopen.95691
Edited by Vilmar Steffen

Contributors
Fadl A. Essa, Abdelfatah Marni Sandid, Taieb Nehari, Driss Nehari, Yasser Elhenawy, Zafar Abbas, Nasir Hayat, Anwar Khan, Muhammad Irfan, Mustakeem Mustakeem, Noreddine Ghaffour, Muhammad Saqib Saqib Nawaz, Sofiane Soukane, Abubakar Sadiq Isah, Balbir Singh Mahinder Singh, Husna Takaijudin, Bao The The Nguyen, Lucio Cardozo-Filho, Guilherme Machado, Donato Aranda, Marcelo Castier, Vilmar Steffen, Vladimir Cabral, Monique Dos Santos, Fábio Nishiyama, Chandra Shekar Besta, Shirish Prakash Prakash Bandsode, Francisco J. Sanchez-Ruiz, Rajeev Kumar Dohare, Parvez Amjad Ali Ansari

Notice
Statements and opinions expressed in the chapters are these of the individual contributors and not necessarily those of the editors or publisher. No responsibility is accepted for the accuracy of information contained in the published chapters. The publisher assumes no responsibility for any damage or injury to persons or property arising out of the use of any materials, instructions, methods or ideas contained in the book.

First published in London, United Kingdom, 2022 by IntechOpen
IntechOpen is the global imprint of INTECHOPEN LIMITED, registered in England and Wales, registration number: 11086078, 5 Princes Gate Court, London, SW7 2QJ, United Kingdom
Printed in Croatia

British Library Cataloguing-in-Publication Data
A catalogue record for this book is available from the British Library

Additional hard and PDF copies can be obtained from orders@intechopen.com

Distillation Processes - From Solar and Membrane Distillation to Reactive Distillation Modelling, Simulation and Optimization
Edited by Vilmar Steffen
p. cm.
Print ISBN 978-1-83962-807-8
Online ISBN 978-1-83962-808-5
eBook (PDF) ISBN 978-1-83962-809-2

We are IntechOpen,
the world's leading publisher of
Open Access books
Built by scientists, for scientists

5,900+

Open access books available

146,000+

International authors and editors

185M+

Downloads

Our authors are among the

156

Countries delivered to

Top 1%

most cited scientists

12.2%

Contributors from top 500 universities

Selection of our books indexed in the Book Citation Index (BKCI)
in Web of Science Core Collection™

Interested in publishing with us?
Contact book.department@intechopen.com

Numbers displayed above are based on latest data collected.
For more information visit www.intechopen.com

Meet the editor

Professor Vilmar Steffen received a BSc and MSc in Chemical Engineering from the Western Paraná State University – Unioeste in 2007 and 2010, respectively. He also completed six months of a postdoctoral research fellowship at the same university in 2015. He obtained a Ph.D. in Chemical Engineering from the State University of Maringá in 2014. Since February 2015, Dr. Steffen has been an assistant professor in the Academic Department of Engineering (DAENG), Federal Technological University of Paraná (UTFPR), Câmpus Francisco Beltrão, Brazil. His research work focuses on process modeling, simulation, and optimization.

Contents

Preface

The distillation process has many applications, the most important of which is purifying water.

Water is one of the most significant resources in the world. Freshwater is of particular importance because it is increasingly being polluted, threatening human and animal health. One of the most common and effective techniques to purify and maintain the supply of freshwater is distillation, which separates dissolved solids or excess salt, other minerals, and impurities from seawater. Like distillation, membrane distillation uses a hydrophobic membrane between the feed and the condenser, allowing only solvent vapor to pass through. Thus, it has some advantages over conventional desalination by distillation. In the first section, this book presents some principles and studies of desalination by solar and membrane distillation.

Reactive distillation is a process in which chemical reactions and separation by distillation occur in a single operation. Reactive distillation can be used, for example, to separate azeotropic mixtures in equilibrium-limited reactions. Other advantages of reactive distillation are the reduction of investment and operational costs, increased efficiency of the separation process efficiency and improvement of reaction selectivity Reactive distillation has been studied a lot in the last decades and researchers have proposed some modifications to the process, such as adding a reactive divided wall distillation column. In the second section, this book discusses reactive distillation.

A good way to better understand a process is to represent the process with a mathematical model based on conservation laws (mass, energy, and momentum), constitutive relationships, and empirical equations. The proposed mathematical model must be validated by experimental data to prove that it accurately represents the process behavior. Once there is a good mathematical model, it is possible to run simulations. After obtaining a good representation of process behavior, an optimization step can be carried out. This book presents some cases of mathematical modeling, simulation, and optimization of the distillation process. It also includes studies on the application of artificial neural networks in desalination and reactive distillation.

Vilmar Steffen
Departamento acadêmico de Engenharias,
Universidade Tenológica Federal do Paraná,
Francisco Beltrão, Paraná, Brazil

Desalination by Solar and Membrane Distillation

Chapter 1

Principles and Modes of Distillation in Desalination Process

Abubakar Sadiq Isah, Husna Takaijudin
and Balbir Singh Mahinder Singh

Abstract

Distillation has been a very important separation technique used over many centuries. This technique is diverse and applicable in different fields and for different substances. Distillation is important in the desalination section. Various principles are used in desalting seawater and brackish water to fulfill the demands of freshwater. This work explains the modes and principles of distillation in desalination, their types, present improvement, challenges, and limitations as well as possible future improvements. The first and primary mode of distillation is the passive type. As times went by and the demand for freshwater kept increasing, other modes were introduced and these modes fall under the active distillation type. However, each mode has its own advantages, disadvantages, and limitations over each other. The principles and modes of distillation are as significant as understanding the energy sources needed for distillation. Hence, they are the basic knowledge needed for future innovation in the desalination industries.

Keywords: history of distillation, desalination, renewable and nonrenewable energy sources, modes of distillation, principles of distillation

1. Introduction

Over the ages, the world has been evolving in development and resources use, and this has led to enormous waste generation of different states (solid, liquid, and gas). The waste needs to be either treated or recycled, paving ways for different techniques for different wastes to be treated or recycled. One of the important resources on earth is water. It is used for everyday activities such as domestic, industrial, and commercial purposes. This has caused reduction in freshwater quantity globally and shortage in clean water supply because of pollution of the existing sources. Hence, different techniques and approaches are still being investigated that can provide adequate and sustainable freshwater. Distillation has been a promising process of separating components by heating/boiling, which causes evaporation, and cooling, which causes condensation. Distillation is a simple technique of converting liquid to vapor by heating and subsequently condensing it back to liquid after the vapor comes in contact with a cooler surface. Simple distillation may not be efficient for certain modes of treatment; therefore, some other advanced distillations were found like the fractional distillation for petroleum refining and multi-effect distillation (MED) for desalination. Generally, distillation is meant to separate a homogenous fluid mixture using the differences in the volatility or boiling point of the mixture's components [1].

There are three definitions of distillation relevant to desalination. (a) Distillation is a process in which a liquid sample is volatilized into vapor that is later condensed into liquid with richer volatile components of the original sample. This can be achieved by heating, reducing pressure, or both. (b) Distillation is the process of separating a mixture of fluids using the differences in their boiling point or relative volatility. (c) Distillation is the application of heat to a liquid to cause its partial vaporization, and then, a separate vessel is used to collect the condensed vapor [2].

The cost for all distillation methods varies, but they have a similar process or working principle. The temperature difference allows water to evaporate even at 40°C leaving the dissolved solids behind, which require about 300°C to volatilize [3].

Distillation has various advantages such as (i) the capacity to take care of a wide range of feed flow rate range, meaning they can handle high and low flow rates contrary to some alternative techniques. For example, facultative, stabilization, oxidation, and maturation ponds all require a high flow rate of feed; (ii) it can remove various and lots of substances from feed concentrations. Numerous alternative treatments have different stages or include varied chemicals for a particular impurity removal. For example, alum is used mainly to reduce solids through coagulation and chlorine is used only for the elimination of pathogens; so, it cannot remove suspended solids or other impurities; (iii) it can produce water of very high quality (pure); this is contrary to other techniques that partially treat or only reduce the impurity level of the feed. Distillation is a very well-known technique for purification because of its robustness and versatility [1]. One of the major issues with distillation in desalination is the high energy demand for the process. **Figure 1** shows a representation of the distillation process in desalination. After feedwater is transferred to the basin, the first step is the use of energy, mostly solar energy, to heat the basin water to cause it to evaporate to produce freshwater; the byproduct remains in the basin as brine solution, which can also be extracted.

The aim of this chapter is to elaborate the principles and modes of distillation in desalination and analyze their types, improvements, features, challenges, limitation, cost, gap, and future improvements needed.

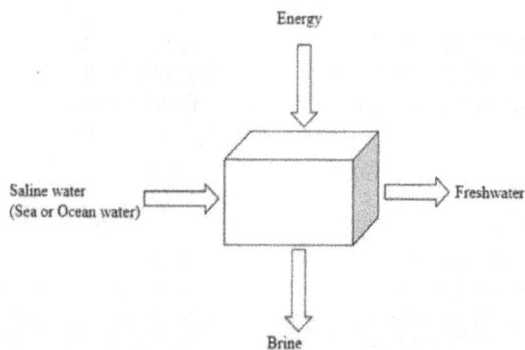

Figure 1.
Distillation process in desalination.

2. History of distillation in desalination

Despite distillation being widely used in various disciplines lately, it was first used for desalination by the people of Babylonia in Mesopotamia, which was found on the Akkadian tablet dated c. 1200 BCE. Later, Aristotle (384–322 BC) established a hypothesis that when saltwater evaporates, it forms vapor, which becomes sweet, and the condensate is salt free. Pliny the elder (AD 23–70) explained on the purification of seawater, specifically the Red Seawater *via* pearl barley leaves, the leaves absorb the salt content in the seawater. The leaves are spread around the ship so that they can absorb seawater and this makes the leaves moist; then the clean water is extracted by squeezing. Alexandria the chemist in Roman Egypt during the first century narrated on how sailors used bronze vessels covered with sponges to boil seawater and how condensates are collected by the sponges [4].

Furthermore, evidence of baked clay retorts and receivers was found at old Indian subcontinent cities; cities such as Taxila, Charsadda, and Shaikan Dheri in modern Pakistan show evidence that during early centuries distillation was practiced there. The distillers were locally called Gandhara stills and they could only produce weak liquor because they lacked efficient means for vapor collection at low heat. However, the first distinct use of distillation specifically for water (distill water) was in 200 CE by Alexander of Aphrodisia. The process continued for other liquids in the early Byzantine Egyptian during the third century under Zosimus of Panopolis [5].

In the eighth and ninth centuries, wine distillation was attributed to Arabic work by Al-Kindi and Al-Farabi, and some were found in the 28th book of Al-Zahrawi commonly known as Abulcasis. During the centuries mentioned earlier, some Medieval chemists such as Jabir ibn Hayyan known as Geber and Abu Bakr al- Razi known as Rhazes did rigorous experiments on distillation using various substances. Later in the twelfth century, a popular recipe known as aqua ardens, which means burning water, which in turn means ethanol, was produced by distilling wine with salt and by the end of the thirteenth century, it became very common in the Western European chemists [5].

In China, distillation started during Eastern Han Dynasty between the first and the second centuries, then in Southern Song between the tenth and thirteenth centuries from archeological findings, and then later in Jin between the twelfth and thirteenth centuries, although the process was predominantly related to the distillation of beverages. In the thirteenth and fourteenth centuries in Qinglong, Hebei Province of China, distillation of beverages was common during the Yaun Dynasty [4, 5].

The trend continued and up to 1500 and a German alchemist Hieronymus Braunschweig published a book called "The book of the Art of Distillation." This was the first book on distillation and in 1512, the scope was expanded. In 1651, a book titled "Art of Distillation" was published by John French even though most of the work was from Hieronmus [5].

Alchemy later evolved into the science of chemistry, and local equipment such as alembic and retorts now became vessels or glassware in general terms. Until recently, some of the equipment like pot still made of different materials are still used for domestic production or in the manufacture of essential oils [4, 5].

In the modern or middle civilization, that is, during 1822, Anthony Perrier developed continuous still, which was later improved by Robert Stein in 1826. Aeneas Coffey further improved the still in 1830. His unit is referred to as the archetype of modern petrochemical unit. Ernest Solvay was the first to develop a distillation unit that specifically targeted ammonia removal (ammonia distillation) [5].

Currently in the twenty-first century, from the knowledge of the predecessors, various modifications were made to enhance the yield of the distillate. This led to the development of different types of desalination systems and an increase in their usage, especially to meet the need of providing water for workers on the sea or mining regions [6].

2.1 Distillation desalination

Distillation in water desalination is a technique Or excess salts from saline water. Other minerals and impurities are from seawater or brackish water also removed during desalination and this treatment process can be extended to wastewater, industrial water, rivers, streams, lake, pond, and groundwater/wells. These salts and minerals occurred because of salts. Two products are obtained after desalination—freshwater and brine, which is the waste or byproduct [7].

Desalination can alleviate the pressure on water resources and has the capacity to provide adequate clean water especially to coastal regions and is increasingly becoming an alternative for domestic and industrial freshwater supply. Desalination requires a large amount of energy; however, various energy types can be used for desalination, which makes it a good alternative. **Figure 2** shows the different energy sources that can be used for desalination. They are categorized into nonrenewable energy sources, which include nuclear, coal, petroleum, natural gas, and hydrocarbons, while the renewable energy sources include wind, geothermal, solar, and biomass. The nonrenewable sources are sometimes expensive or in some cases, not environmentally friendly. On the other hand, renewable energy sources such as solar, wind, and geothermal can replace the renewable energy and are abundant and cost efficient to harness, particularly, solar energy that can be used even in rural areas [7].

Globally, there are about 21,000 desalination plants, particularly in Saudi Arabia, United Arab Emirates, and Israel [8]. In the desalination field, distillation can occur as membrane desalination and nonmembrane (thermal) desalination.

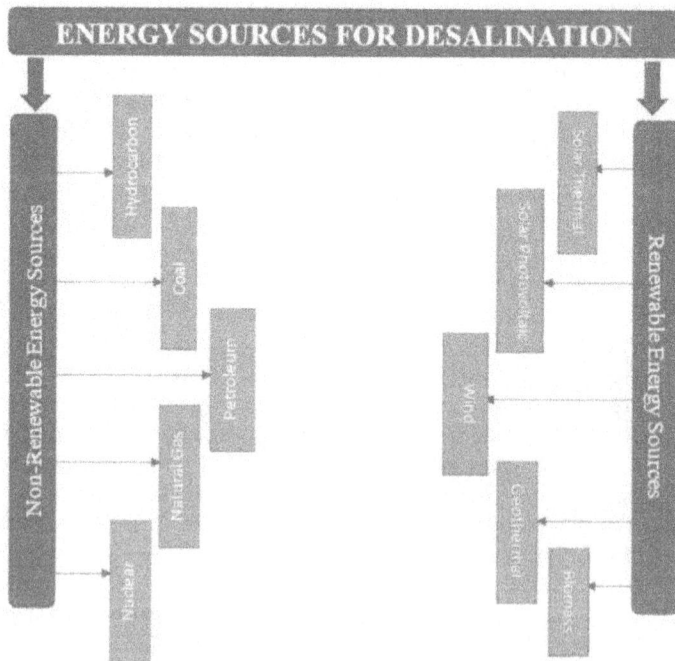

Figure 2.
Energy sources for desalination.

The membrane desalination is the type that is not a complete thermal process; that is, a membrane is needed to complete the process unlike the thermal (non-membrane) process, which does not require such medium but undergoes complete thermal process. The membrane is a porous material with a thin film, which allows water molecules to pass through, while at the same time preventing salts, larger molecule, pathogens, and metals to pass through. The most common type of distillation in desalination is the membrane distillation. Membrane distillation majorly targets seawater and brackish water [9].

The membrane desalination process includes electrodialysis (ED) and reverse osmosis (RO), which are two major desalinations used recently. They are reverse osmosis and thermal desalination systems, which account for 63.7% and 34.2% of total capacity produced, respectively. The thermal desalination includes multi-effect desalination, multi-stage flash (MSF) desalination, humidification-dehumidification, vapor compression desalination (VCD), and solar still [9].

2.2 Principles of distillation in desalination

Distillation is an ancient method of desalination. It is a phase change process where the liquid known as feedwater, which is mostly seawater or brackish water, is heated to the gaseous state known as vapor and then condensed back to liquid. The condensed water is separated leaving behind brine (byproduct) during the process of evaporation and condensation. There are different distillation types in desalination, namely, solar distillation, multi-effect distillation, multi-stage flash distillation, vapor compression distillation, and membrane distillation.

2.2.1 Solar distillation

Solar distillation imitates the natural hydrological cycle in which solar energy heats the water, causes it to evaporate, and the vapor upon encountering cool surface condenses (**Figure 3**). The condensate is mostly referred to as distillate, which is the freshwater produced, while the impurities left behind is called the brine, which is the byproduct [8]. The first solar distiller was built by Carlos Wilson in Las Salinas in Chile in the year 1872. The distillation principle in this method is that the sun heats the feedwater in the basin and the water molecule evaporates. When the evaporated water molecule (vapor) touches the still cover, which is usually cooler than the vapor, it then condenses to form droplets on the cover. The droplets keep

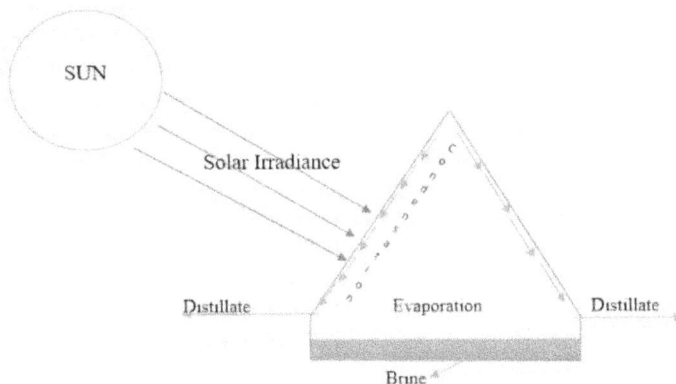

Figure 3.
Schematic diagram of solar distillation.

increasing in size until they reach a size that they can slide down *via* the cover and through the channel for collection. The brine remains in the basin [3]. The parts of a solar distiller are a glass cover and a basin.

The major advantage of solar distillation is the free energy sources, which is the solar energy. There are other numerous advantages of this process such as design simplicity, low cost of fabrication, and maintenance. However, the major disadvantage of this process is the limitation of the sun at night and during cloudy or rainy times. The scale can easily corrode the basin as well. Sometimes, they do not adequately treat nutrient pollutants. In addition, the distillation rate is slow, and the yield is usually small in quantity compared to the other techniques. The average volume of water produced from conventional solar still is 0.8 liters per hour of sun per meter square [10]. In 2014, globally the cost of freshwater from solar distiller ranged between 0.019\$/m^3 and 0.02\$/m^3 depending on the shape of the still [11].

2.2.2 Multi-stage flash distillation

This process is like a continuous process for solar distillation. In this process, the feedwater is first pretreated; it then gets heated and evaporated in the first chamber or stage and the released energy from the condensation is used to heat the water in the second stage and continuously to the last stage after which post-treatment occurs and freshwater is obtained (**Figure 4**). This means that each flash process uses the energy from the previous vapor [8]. The process has several series of flash chambers. Unlike multi-effect distillation in multi-stage flash distillation, heating and boiling occur in the same vessel. The estimated unit cost of freshwater produced from MSF is 1.40\$/m^3 as of 2018 [12].

The advantage of this system is that it minimizes the operating cost because the heat released from each stage is being reused (waste heat). The second advantage is that the strength of the feedwater does not really affect the overall freshwater quality produced because of multiple distillation process for each chamber. Finally, a large quantity of freshwater is produced. The disadvantage of this process is the scale formation during heating, although the scale remains in the brine rather than the heating surface which majorly increases the maintenance cost and frequency but do not damage the system [3]. Features of MSF include Stages (spaces), heat exchanger, distillate collector, and brine heater.

2.2.3 Multi-effect distillation

The multi-effect distillation process involves spraying the feedwater on the pipe to heat the feedwater and generate steam. The steam is utilized to heat the subsequent feedwater and evaporate it to produce freshwater and brine as

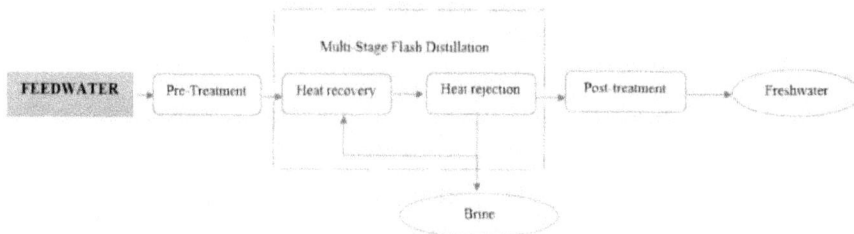

Figure 4.
Schematic diagram of multi-stage flash distillation.

Feedwater   **Feedwater**

Vapour

Hot heat
Transfer fluid   Vapour

Membrane

Cold heat   Brine   Distillate
Transfer fluid   Brine

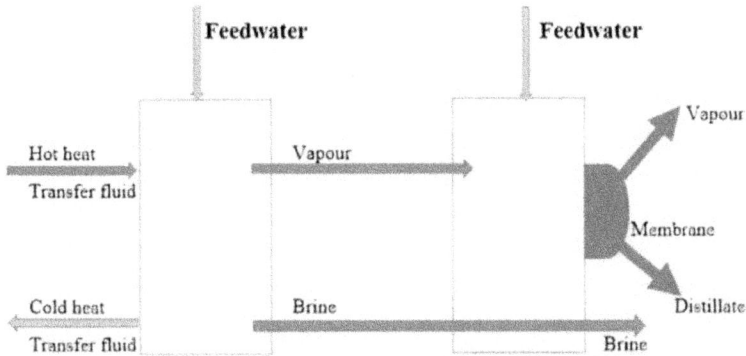

Figure 5.
Schematic diagram of MED in two stages.

byproduct (**Figure 5**). The energy is obtained through a solar collector. Flat plat collector and evacuated tube collector are energy sources for small-scale MED, while parabolic trough collector or any collector that concentrates solar energy is used for large scale. MED is a very ancient process, and only at the first stage, the first steam is independently generated. Subsequent stages use the vapor from the first and previous stage as energy source. There are about 8 to 16 effects for most MED. More number of effects means more efficiency [13]. The goal of MED is to use same heat to evaporate more feedwater. That is, the heat from the first stage helps in evaporation in the second stage and the heat from the second stage aids evaporation in the third. At the same time, each stage evaporator acts as a condenser for each previous stage. This way, large latent heat of vaporization is reused several times before dissipating to the surrounding, but it is significant that the temperature of the first effect is lower than the boiler heating steam [3]. These energy sources when tapped from the sun are converted to electrical energy to provide heat for the pump.

The advantages of MED include low consumption of energy in comparison with other thermal techniques; it works at low temperature and concentration to minimize scaling and corrosion. Pretreatment is not essential. It is very reliable and have low maintenance costs. The disadvantage of this distillation process is that there is heat and pressure losses at each stage because the process is not adiabatic and this can reduce the freshwater yield. There is corrosion and erosion at the contact surfaces between the brine and heat exchanging surface [13]. In 2003, it was found out that the average cost of freshwater worldwide that is produced from MED is 1.00$/m³ which is lower than MSF [12]. Features of MSF include heat source, heat sink, stage and distillate collector, and a membrane (**Figure 5**).

2.2.4 Vapor compression distillation

This process requires a jet stream or mechanical compressor to compress the vapor above the liquid unlike MED and MSF that require energy sources such as crude oil, wind, natural gas, and so on. The compressed vapor supplies heat to the rest of the feedwater for evaporation. Even though the process is a complex type and is mostly used for small-scale distillation, it is far more effective than MED because one effect of VCD is almost as effective as 15 to 20 effects of MED. In **Figure 6**, the feedwater is preheated in the heat exchanger. Later, it is transferred to the tube of the evaporator where it is boiled, and the vapor goes to the mechanical compressor.

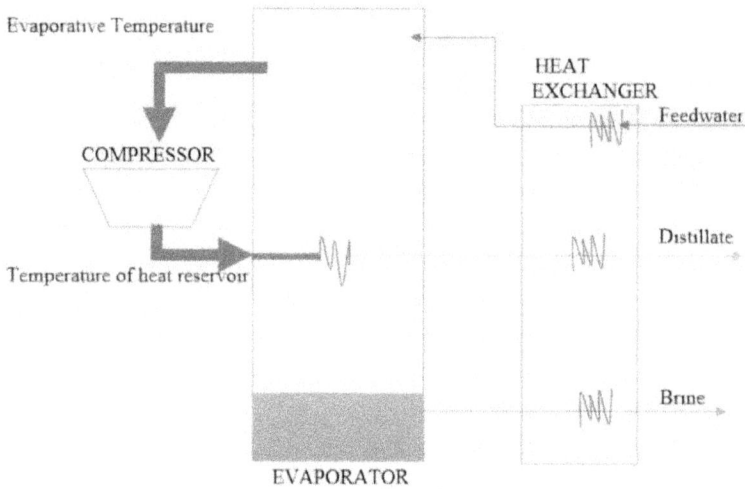

Figure 6.
Schematic diagram of VCD.

The vapor is compressed by the mechanical compressor. The hot compressed water vapor is transferred back to the evaporator, which is condensed outside the tube at the same time supplying the heating energy required for boiling feedwater. The non-condensing gases are removed with the help of a vacuum pump or ejector [3].

The advantages of this method are low operating and maintenance cost; it has a vast temperature range for operation; it is very efficient and reliable because it has good water recovery ratio and moderate energy consumption; and it is easy to use and maintain. The huge initial cost is a major disadvantage for this process. Second, it requires pretreatment to minimize fouling and scaling, and internal scaling can occur as a result of crystals accumulation in the pore [14]. Finally, quality materials are needed to prevent corrosion [15]. The average cost of freshwater produced from vapor compression distillation is 0.93$/m^3 as at 2009 findings [16].

2.2.5 Membrane distillation

This process uses differences in temperature across the membrane to evaporate the feedwater and condensed the freshwater leaving the impurities, salts, and other minerals in the form of brine solution. The concept of membrane distillation is microfiltration, which allows only water molecules to pass through porous hydro-phobic membrane [17]. Membrane distillation can use different low-grade energy

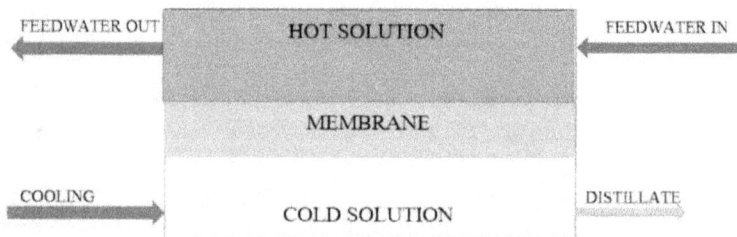

Figure 7.
Schematic diagram of the direct contact membrane.

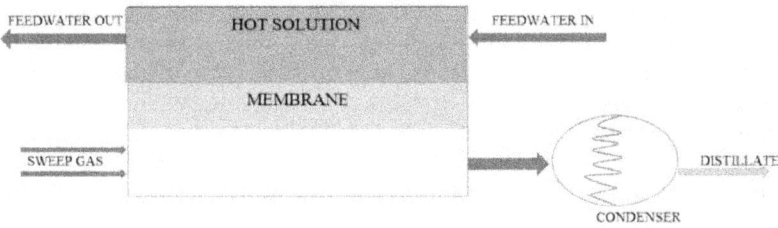

Figure 8.
Schematic diagram of the sweep gas membrane.

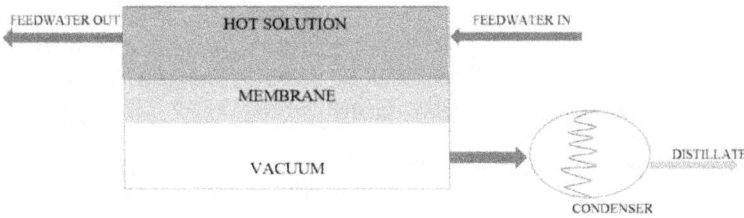

Figure 9.
Schematic diagram of the vacuum membrane.

sources like the sun or wind. The water molecules move from the region of high to low vapor pressure through the membrane. There are four methods by which the vapor is recovered through the membrane. The first is through the direct contact of the liquid phase with both sides of the membrane to obtain distillate and the condensation process is controlled by the thickness of the membrane (**Figure 7**). Although the heat loss in this method is higher than that in other methods because

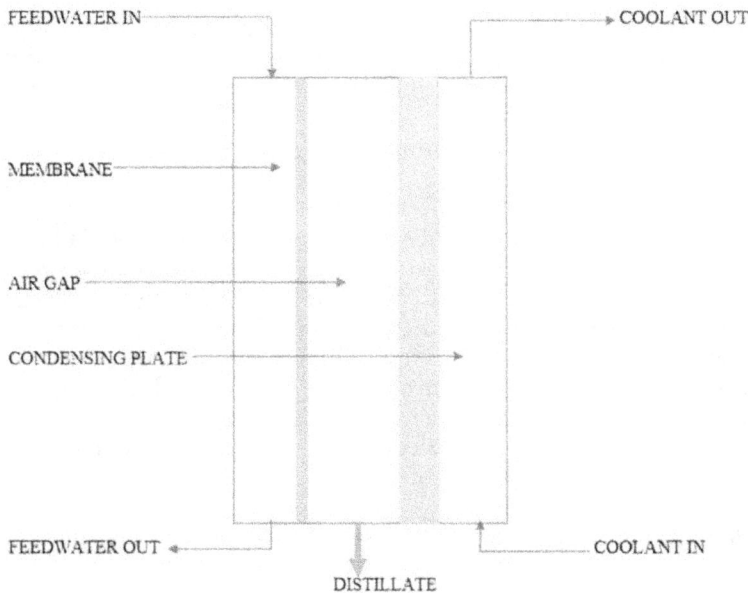

Figure 10.
Schematic diagram of air gap membrane.

of continuous contact between the membrane, the hot feed and cold permeate. This method is called the direct contact membrane distillation. The second method is vapor withdrawal by using a vacuum on the permeate region; in this method, the process is like the first but for the introduction of the condenser and the sweep gas that differentiate them (**Figure 8**). This is also known as the sweep gas membrane distillation. The third method is having an external condensation and in this case, the vapor is removed by using an inert gas stream (**Figure 9**). This method is called vacuum membrane distillation [18]. The fourth method is the addition of air gap interposed between the condensation surface and the membrane (**Figure 10**). This process is called the air gap membrane distillation [19].

The advantages of membrane distillation include a high-rate removal of macromolecules and other substances, lower operating temperature and pressure, unadulterated interaction between the membrane and the process, and reduction in vapor spaces. The main disadvantage of membrane distillation is membrane wetting, which is caused as a result of fouling and excessive liquid entry pressure [18]. In 2004, the average cost of freshwater from membrane distillation was 0.705$/m^3 [20].

2.3 Simple desalination system

This section will focus on the first concept of distillation in desalination, improvements made on them till date, the merits, and their demerits as well as the challenges and limitations of all the mentioned distillation desalination systems experienced so far. The first developed distillation system is known as the conventional desalination system. The first step is fabrication of a basin, which is usually made of metal material. Then, the fabrication of the cover that is mainly glass material was used. The cover generally is in a triangular shape. Afterward, the glass cover is placed on the basin to form a closed system that will allow distillation/desalination. In ancient times, they did not know how to seal the bottom edges between the glass cover and the basin because of which there were high heat losses and a low yield. Later, the system was sealed with mostly silicone gel to prevent or minimize heat loss and increase the yield. This is a simple way for conventional desalination that can be replicated anywhere in the world. **Table 1** shows the different types of distillation systems, the improvisation made, current challenges, and limitations, with possible future improvements.

2.4 Modes of distillation in desalination

Distillation in desalination can occur in three different modes, although the third mode is a combination of the two independent modes. The first is the utilization of the solar irradiance directly from the sun causing heating/evaporation to the feedwater and condensation when it meets a cold surface. The second mode is the storage of the solar irradiance or any other sources of energy such as wind and geothermal to produce electricity that is used to heat and evaporate the feedwater and later condensed to obtain the distillate [18]. The third is the utilization of the solar irradiance directly and supplementing it with other energy sources that produce electricity or other forms of heating source to cause distillation.

2.4.1 Passive distillation

Passive desalination or distillation is the cheapest and most used method even in rural regions. This is because of its simplicity and because it can work on its own. It uses only energy from the sun (solar irradiance) directly to heat the water leading to

Distillation desalination types	References	Present improvements	Challenges/limits	Gap/feature improvement
Solar Distillation	[21–28]	• Geometry improvement such as cover angle, shape, thickness. The cover angle ranges from 2 mm to 7 mm. The angle ranges from 10^0 to 60^0. The shapes include triangular, tubular, hemispherical, trapezoidal, and circular. These improvements are to minimize heat loss. • Use of different heat absorbers which include paraffin wax, sand, black stone, sponge, and wick to increase heat retention. • Use of nanoparticle such as Al_2O_3 nanoparticle, and SiC nanofluids to fastened evaporation. • Use of different sealants such as sawdust, glass wool, ply wool hey, and silicone gel to minimize heat loss. • Use of photovoltaic to augment solar irradiance.	• Low yield. • The conventional system cannot be used during rain, cloud, or at night because of the absence of sunlight. • Cost of solar photovoltaic can be overwhelming and limited energy storage capacity. • Some solar distillers cannot completely remove nutrient if it is in high concentration in the feedwater.	• More heat-retaining substances that can be used for the basin. • More heat-absorbing substance that can be used for the cover. • Advance usage of other energy sources such as biomass, wind to augment solar radiation.
Multi-stage flash (MSF)	[29–36]	• Increasing the number of brine heater to increase the evaporation rate. • Some materials such as aluminum, brass, or titanium tubing for evaporators, combination of Cu/Ni for the evaporator wall to enhance water production. • Use of certain energy sources such as super-critical carbon dioxide Brayton cycle, solar tower power, and parabolic trough solar-based power to increase energy harnessing. • Coproduction of electricity and water from steam turbines. • Use of solar steam as heat source for brine heater.	• Scale formation and corrosion. • Hot distillate. • High capital as cannot be used for small scale. • The need to pretreat feedwater to overcome thermal desalination problem. • System failure because of complexity.	• Use of alternative materials or locally made materials to reduce cost. • Reduction in scale formed through filtration membrane to increase plant efficiency.

Distillation desalination types	References	Present improvements	Challenges/limits	Gap/feature improvement
Multi-effect distillation	[29, 37–42]	• Use of air-cooled condensers. • Integrating MED and other desalination types such as membrane, adsorption cycle, and reverse osmosis to enhance freshwater recovery and reduce brine volume. • Use of sprays to reduce internal heat loss. • To reduce corrosion and increase system performance, special metal alloys, or polymers are used for the MED evaporator. • Use of equipment with high temperature. • Permeate reprocessing in different configuration. • Combination of MED with gas turbine plant.	• More surface area may be needed for recirculation of energy. • Not suitable for small scale, because of high capital cost. • The special materials used for MED evaporator could be very costly and some have low thermal conductivity. • System complexity.	• Improving hybrid systems (combining different system) for better performance. • Use of multiple energy sources. • Air cooled condensers specially made from nanoparticle with improved cooling effect. • Special nanoparticle materials may be used to fabricate evaporator.
Vapor compression	[43-52]	• Use of substances such as polymer heat transfer elements to enhance freshwater production. • Use of high-temperature heat pumps to increase energy efficiency for VC as well as choosing a better working fluid like ammonia, carbon dioxide, hydrocarbons, and synthetic working fluid or their combination. • Optimizing cycle configuration into different series to giver better system performance. • Making it a hybrid system like adding internal solution circuit. • Hybrid systems for multipurpose use by treating brine and recovering salts and minerals.	• High energy demand and the need to investigate to use cheaper energy technology. • High initial/capital cost. • Contamination of the vapor water with some vapor compressions system material such as SO2, NO and even H$_2$SO$_4$.	• New different working fluids can be experimented to check their suitability for VC. • Investigate materials suitable for VC with little or no contamination.

Distillation desalination types	References	Present improvements	Challenges/limits	Gap/feature improvement
Membrane	[53–62]	• Use of carbon nanotube like polydimethylsiloxane in different format to increase system efficiency. • Use of thin film nanocomposite/nanomaterials like carbon to fabricate membrane systems to improve system efficiency. • Use of interfacial polymers to enable better heat recovery. • The use of different synthetic membranes. • The use of materials with anti-wetting properties like amphiphobic polyvinylidene fluoride co-hexaflouropropylene.	• One of the major challenges of membrane desalination is the high osmotic pressure generated because of high ionic concentrations of saline water in addition to the fouling problem. • Not reliable because of limitation of sun unless solar energy is harvested.	• More nanomaterials like Al_2O_3 can be investigated to check their suitability in improving membrane desalination. • Hybrid or integration of other approaches with membrane desalination can be experimented to find out their efficiency.

Table 1.
Challenges, present improvements, and future prospective of distillation desalination types.

its evaporation, and the vapor condenses when it touches the cold surface. For this type of distillation to occur, an enclosed system is needed, which basically consists of a water basin and a transparent cover usually made of glass or plastic. It is then sealed to prevent vapor or heat loss [2].

The advantage of this process is that it has a cheap source of energy, as the energy comes directly from the sun. This energy is abundant in most regions and does not need conversion or storage. It is simple to fabricate, use, and maintain. It can produce very clean water for drinking and other purposes without the need for further treatment. However, the major disadvantage of this process is that it can operate only during the day when there is sunshine. At nighttime or rainy times, since there is no sun, it is not possible to carry out this process. Solar stills, solar chimneys, and humidification-dehumidification are examples of this process.

2.4.2 Active distillation

In active distillation, the same process is observed. In other words, there is heating, evaporation, and condensation of feedwater to obtain freshwater. It is just that in this case, the sources of energy are other sources such as wind, geothermal, or stored solar energy in photovoltaic cells, which are later converted to alternative current to supply the energy/heat for the distillation process.

Some of the advantages of this process include faster distillation, as the feedwater gets heated faster than it does in the passive method. The method can also be used either nighttime or daytime, especially during cloudy or rainy days when there is no solar radiation. Evident research carried out at Univerisiti Teknologi PETRONAS, Malaysia, shows that this method produces cleaner freshwater than passive and combined distillation. This method, however, has some challenges like more scale formation and higher and more expensive energy usage. It may require semi-skilled persons to operate some of these distillers.

Solar stills, solar chimneys, and humidification-dehumidification can also be used for this process. In addition, membrane desalination can also fall in this category, since some part of it involves phase change of the feed (distillation) with the aid of a membrane.

2.4.3 Combined

In the combined state, this is also operated during the daytime, where both solar irradiance and other energy sources such as photovoltaic, wind, and biomass are used simultaneously for distillation. This produces more freshwater than the first two mentioned methods because of the effect of combined efforts. This process increases the frequency of performing maintenance tasks, as corrosion and scale formation is high.

3. Conclusions

Historical evidence has shown that distillation has been an old technique for water purifications. Distillation in desalination has proven to be an effective technique for freshwater production. Various distillation types have a similar quality of distillate. However, they differ in distillate quantity. The mode of distillation can also have an impact on the quality of freshwater yield as evidently carried out in the laboratory where the active mode had cleaner water production. The distillation types in desalination have advantages and disadvantages over one another. Therefore, the quantity and quality of distillate needed determines the most appropriate distillation type to choose for the desalination process. Currently, there are numerous researches that are exploring on how to enhance the distillation process in desalination. There is still the need for technological advancement to enhance yield in the case of solar distillation and reduce cost, especially in the vapor compression desalination process. The carbon nanotube membrane has been a promising solution for membrane desalination and can be exploited further. Overall, desalination has been an effective and efficient solution to augment conventional clean water supply. If given proper attention, it can be a lasting solution for clean water supply.

Acknowledgements

The author wishes to thank Universiti Teknologi PETRONAS Malaysia in collaboration with Ahmadu Bello University, Zaria for providing an enabling environment and support. The author also wishes to appreciate the YUTP-MPSS with cost centers 015LC0-215.

Conflict of interest

The authors declare no conflict of interest.

Notes/thanks/other declarations

Special thanks to Dr. Husna Takaijudin, APDR Balbir Singh Mahinder Singh, and APDR Kamaruzaman Wan Yusof for their mentorship and guidance.

Appendices and nomenclature

ED	Electrodialysis
RO	Reverse osmosis
MSF	Multi-Stage flash distillation
MED	Multi-effect distillation
VCD	Vapor-compression distillation

Author details

Abubakar Sadiq Isah*, Husna Takaijudin and Balbir Singh Mahinder Singh
Universiti Teknologi PETRONAS, Perak, Malaysia

*Address all correspondence to: sadiqisah191@gmail.com

IntechOpen

References

[1] R. Smith and M. Jobson, "Distillation," Principles and Practice Modern Chromatographics Methods. Elsevier, 9, 2019. Available: https://www.sciencedirect.com/topics/chemistry/distillation

[2] Green JD. Distillation. Chemical Engineering Journal. Elsevier. 2017 Available: https://www.sciencedirect.com/topics/earth-and-planetary-sciences/distillation

[3] Saidur R, Elcevvadi ET, Mekhilef S, Safari A, Mohammed HA. An overview of different distillation methods for small scale applications. Renewable & Sustainable Energy Reviews. 2011;**15**(9):4756-4764. DOI: 10.1016/j.rser.2011.07.077

[4] Birkett JD. The history of Desalination Before Large-Scale Use. In: Encyclopedia of Desalination and Water Resources. Vol. I. History, Development and Management of Water Resources; 2000. p. 9

[5] Wikipedia, "Distillation," *wikipedia*. 2021, Available: https://en.wikipedia.org/wiki/Distillation

[6] Campero C, Harris LM. The legal geographies of water claims: Seawater desalination in mining regions in Chile. Water (Switzerland). 2019;**11**(5):1-21. DOI: 10.3390/w11050886

[7] J. Han, A. Paytan, P. Shi, A. Huang, "Desalination," *wikipedia*. Elsevier, 2021, Available: https://www.sciencedirect.com/topics/earth-and-planetary-sciences/desalination

[8] Wikipedia, "Desalination." 2021, Available: https://en.wikipedia.org/wiki/Desalination

[9] V. Mohammad-Razdari and S. A. Fanaee, "Comprehensive review of different types of desalination," Journal of Renewable and New Energy, 72, 2-4, 2020, Available: https://dorl.net/dor/20.1001.1.24234931.1400.8.1.3.7

[10] Singh SK, Kaushik SC, Tyagi VV, Tyagi SK. Comparative Performance and parametric study of solar still: A review. Sustainable Energy Technologies and Assessments. 2021;**47**:101541. DOI: 10.1016/j.seta.2021.101541

[11] Ranjan KR, Kaushik SC. Economic feasibility evaluation of solar distillation systems based on the equivalent cost of environmental degradation and high-grade energy savings. International Journal of Low-Carbon Technologies. 2014;**11**(1):8-15. DOI: 10.1093/ijlct/ctt048

[12] Van Der Bruggen B. Desalination by distillation and by reverse osmosis - Trends towards the future. Membrane Technology. 2003;**2003**(2):6-9. DOI: 10.1016/s0958-2118(03)02018-4

[13] Ghenai C, Kabakebji D, Douba I, Yassin A. Performance analysis and optimization of hybrid multi-effect distillation adsorption desalination system powered with solar thermal energy for high salinity sea water. Energy. 2021;**215**:119212. DOI: 10.1016/j.energy.2020.119212

[14] Chang YS, Ooi BS, Ahmad AL, Leo CP, Low SC. Vacuum membrane distillation for desalination: Scaling phenomena of brackish water at elevated temperature. Separation and Purification Technology. 2021;**254**:117572. DOI: 10.1016/j.seppur.2020.117572

[15] Lara JR, Noyes G, Holtzapple MT. An investigation of high operating temperatures in mechanical vapor-compression desalination. Desalination. 2008;**227**(1-3):217-232. DOI: 10.1016/j.desal.2007.06.027

[16] RAHMAH ML. Desalination Using Vapor-Compression Distillation. Texas A&M University; 2009

[17] Alkhudhiri A, Darwish N, Hilal N. Membrane distillation: A comprehensive review. Desalination. 2012;**287**:2-18. DOI: 10.1016/j.desal.2011.08.027

[18] Qtaishat MR, Banat F. Desalination by solar powered membrane distillation systems. Desalination. 2013;**308**:186-197. DOI: 10.1016/j.desal.2012.01.021

[19] Ahmed FE, Lalia BS, Hashaikeh R, Hilal N. Alternative heating techniques in membrane distillation: A review. Desalination. 2020;**496**:114713. DOI: 10.1016/j.desal.2020.114713

[20] Walton J, Lu H, Turner C, Solis S, Hein H. Solar and Waste Heat Desalination by Membrane Distillation. United States Department of the Interior, Bureau of Reclamation. 2004 Available: http://linkinghub.elsevier.com/retrieve/pii/S0376738813007096%0Ahttps://www.usbr.gov/research/dwpr/reportpdfs/report081.pdf

[21] Reif JH, Alhalabi W. Solar-thermal powered desalination: Its significant challenges and potential. Renewable & Sustainable Energy Reviews. 2015;**48**:152-165. DOI: 10.1016/j.rser.2015.03.065

[22] Sellami MH, Belkis T, Aliouar ML, Meddour SD, Bouguettaia H, Loudiyi K. Improvement of solar still performance by covering absorber with blackened layers of sponge. Groundwater for Sustainable Development. 2015, 2017;**5**:111-117. DOI: 10.1016/j.gsd.2017.05.004

[23] Chen W, Zou C, Li X, Li L. Experimental investigation of SiC nanofluids for solar distillation system: Stability, optical properties and thermal conductivity with saline water-based

fluid. International Journal of Heat and Mass Transfer. 2017;**107**:264-270. DOI: 10.1016/j.ijheatmasstransfer.2016.11.048

[24] Zheng Y, Caceres Gonzalez R, Hatzell MC, Hatzell KB. Concentrating solar thermal desalination: Performance limitation analysis and possible pathways for improvement. Applied Thermal Engineering. 2020, 2021;**184**:116292. DOI: 10.1016/j.applthermaleng.2020.116292

[25] Fujiwara M, Kikuchi M. Solar desalination of seawater using double-dye-modified PTFE membrane. Water Research. 2017;**127**:96-103. DOI: 10.1016/j.watres.2017.10.015

[26] Xu J et al. Solar-driven interfacial desalination for simultaneous freshwater and salt generation. Desalination. 2019, 2020;**484**:9. DOI: 10.1016/j.desal.2020.114423

[27] Jang GG et al. Efficient Solar-Thermal Distillation Desalination Device by Light Absorptive Carbon Composite Porous Foam. Global Challenges. 2019;**3**(8):1900003. DOI: 10.1002/gch2.201900003

[28] Yi L et al. Hollow black TiAlOx nanocomposites for solar thermal desalination. Nanoscale. 2019;**11**(20):9958-9968. DOI: 10.1039/c8nr10117e

[29] Hanshik C, Jeong H, Jeong KW, Choi SH. Improved productivity of the MSF (multi-stage flashing) desalination plant by increasing the TBT (top brine temperature). Energy. 2016;**107**:683-692. DOI: 10.1016/j.energy.2016.04.028

[30] Al-Mutaz IS. Msf challenges and survivals. Desalination and Water Treatment. 2020;**177**:14-22. DOI: 10.5004/dwt.2020.24908

[31] Alhazmy MM. Multi stage flash desalination plant with brine-feed mixing and cooling. Energy.

2011;**36**(8):5225-5232. DOI: 10.1016/j. energy.2011.06.024

[32] Darawsheh I, Islam MD, Banat F. Experimental characterization of a solar powered MSF desalination process performance. Thermal Science and Engineering Progress. 2018, 2019;**10**:154-162. DOI: 10.1016/j. tsep.2019.01.018

[33] Moharram NA, Bayoumi S, Hanafy AA, El-Maghlany WM. Hybrid desalination and power generation plant utilizing multi-stage flash and reverse osmosis driven by parabolic trough collectors. Case Studies in Thermal Engineering. 2020, 2021;**23**:100807. DOI: 10.1016/j.csite.2020.100807

[34] El-Ghonemy AMK. Performance test of a sea water multi-stage flash distillation plant: Case study. Alexandria Engineering Journal. 2018;**57**(4): 2401-2413. DOI: 10.1016/j. aej.2017.08.019

[35] Toth AJ. Modelling and optimisation of multi-stage flash distillation and reverse osmosis for desalination of saline process wastewater sources. Membranes (Basel). 2020;**10**(10):1-18. DOI: 10.3390/membranes10100265

[36] Khoshrou I, Jafari Nasr MR, Bakhtari K. New opportunities in mass and energy consumption of the Multi-Stage Flash Distillation type of brackish water desalination process. Solar Energy. 2017;**153**:115-125. DOI: 10.1016/j.solener.2017.05.021

[37] Tahir F et al. Sustainability assessment and techno-economic analysis of thermally enhanced polymer tube for multi-effect distillation (Med) technology. Polymers (Basel). 2021;**13**(5):1-20. DOI: 10.3390/ polym13050681

[38] Chen Q, Burhan M, Kum Ja M, Li Y, Choon Ng K. A spray-assisted multi-effect distillation system driven by

ocean thermocline energy. Energy Conversion and Management. 2021;**245**:114570. DOI: 10.1016/j. enconman.2021.114570

[39] Al-hotmani OMA, Al-Obaidi MA, Patel R, Mujtaba IM. Performance analysis of a hybrid system of multi effect distillation and permeate reprocessing reverse osmosis processes for seawater desalination. Desalination. 2019;**470**:114066. DOI: 10.1016/j. desal.2019.07.006

[40] Xue Y, Du X, Ge Z, Yang L. Study on multi-effect distillation of seawater with low-grade heat utilization of thermal power generating unit. Applied Thermal Engineering. 2017, 2018;**141**:589-599. DOI: 10.1016/j. applthermaleng.2018.05.129

[41] Alhaj M, Hassan A, Darwish M, Al-Ghamdi SG. A techno-economic review of solar-driven multi-effect distillation. Desalination and Water Treatment. 2017;**90**:86-98. DOI: 10.5004/dwt.2017.21297

[42] Ahmadi P, Khanmohammadi S, Musharavati F, Afrand M. Development, evaluation, and multi-objective optimization of a multi-effect desalination unit integrated with a gas turbine plant. Applied Thermal Engineering. 2020;**176**:115414. DOI: 10.1016/j.applthermaleng.2020.115414

[43] Bamigbetan O, Eikevik TM, Neksā P, Bantle M. Review of vapour compression heat pumps for high temperature heating using natural working fluids. International Journal of Refrigeration. 2017;**80**:197-211. DOI: 10.1016/j.ijrefrig.2017.04.021

[44] Catrini P, Cipollina A, Micale G, Piacentino A, Tamburini A. Exergy analysis and thermoeconomic cost accounting of a Combined Heat and Power steam cycle integrated with a Multi Effect Distillation-Thermal Vapour Compression desalination plant.

Energy Conversion and Management. 2017;**149**:950-965. DOI: 10.1016/j. enconman.2017.04.032

[45] El-Dessouky HT, Ettouney HM, Al-Juwayhel F. Multiple effect evaporation-vapour compression desalination processes. Chemical Engineering Research and Design. 2000;**78**(4):662-676. DOI: 10.1205/026387600527626

[46] Zejli D, Ouammi A, Sacile R, Dagdougui H, Elmidaoui A. An optimization model for a mechanical vapor compression desalination plant driven by a wind/PV hybrid system. Applied Energy. 2011;**88**(11):4042-4054. DOI: 10.1016/j. apenergy.2011.04.031

[47] Panagopoulos A. Techno-economic evaluation of a solar multi-effect distillation/thermal vapor compression hybrid system for brine treatment and salt recovery. Chemical Engineering and Processing: Process Intensification. 2019, 2020;**152**:107934. DOI: 10.1016/j. cep.2020.107934

[48] Al-naser KM. Using mechanical vapor compression plant to reduce volume of salts in concentrated liquid. 2019;**1**(5):1-11

[49] Shakib SE, Amidpour M, Esmaieli A, Boghrati M, Ghafurian MM. Various approaches to thermodynamic optimization of a hybrid multi-effect evaporation with thermal vapour compression and reverse osmosis desalination system integrated to a gas turbine power plant. International Journal of Engineering, Transactions B: Applications. 2019;**32**(5):777-789. DOI: 10.5829/ije.2019.32.05b.20

[50] Feria-Díaz JJ, López-Méndez MC, Rodríguez-Miranda JP, Sandoval-Herazo LC, Correa-Mahecha F. Commercial thermal technologies for desalination of water from renewable energies: A state of the

art review. Processes. 2021;**9**(2):1-22. DOI: 10.3390/pr9020262

[51] Samaké O, Galanis N, Sorin M. Thermo-economic analysis of a multiple-effect desalination system with ejector vapour compression. Energy. 2018;**144**:1037-1051. DOI: 10.1016/j. energy.2017.12.112

[52] Buabbas SK, Al-Obaidi MA, Mujtaba IM. A parametric simulation on the effect of the rejected brine temperature on the performance of multieffect distillation with thermal vapour compression desalination process and its environmental impacts. Asia-Pacific Journal of Chemical Engineering. 2020;**15**(6):1-14. DOI: 10.1002/apj.2526

[53] Lu X, Elimelech M. Fabrication of desalination membranes by interfacial polymerization: History, current efforts, and future directions. Chemical Society Reviews. 2021;**50**(11):6290-6307. DOI: 10.1039/d0cs00502a

[54] Kumar M, Khan MA, Arafat HA. Recent Developments in the Rational Fabrication of Thin Film Nanocomposite Membranes for Water Purification and Desalination. ACS Omega. 2020;**5**(8):3792-3800. DOI: 10.1021/acsomega.9b03975

[55] Ali E, Orfi J, AlAnsary H, Lee JG, Alpatova A, Ghaffour N. Integration of multi effect evaporation and membrane distillation desalination processes for enhanced performance and recovery ratios. Desalination. 2020;**493**:114619. DOI: 10.1016/j. desal.2020.114619

[56] Ali S, Rehman SAU, Luan HY, Farid MU, Huang H. Challenges and opportunities in functional carbon nanotubes for membrane-based water treatment and desalination. Science of the Total Environment. 2019;**646**(19):1126-1139. DOI: 10.1016/j. scitotenv.2018.07.348

[57] Huang J, Hu Y, Bai Y, He Y, Zhu J. Novel solar membrane distillation enabled by a PDMS/CNT/PVDF membrane with localized heating. Desalination. 2020;**489**:114529. DOI: 10.1016/j.desal.2020.114529

[58] Christie KSS, Yin Y, Lin S, Tong T. Distinct Behaviors between Gypsum and Silica Scaling in Membrane Distillation. Environmental Science & Technology. 2020. DOI: 10.1021/acs.est.9b06023

[59] An X, Liu Z, Hu Y. Amphiphobic surface modification of electrospun nanofibrous membranes for anti-wetting performance in membrane distillation. Desalination. 2017, 2018;**432**:23-31. DOI: 10.1016/j.desal.2017.12.063

[60] Afsari M, Shon HK, Tijing LD. Janus membranes for membrane distillation: Recent advances and challenges. Advances in Colloid and Interface Science. 2021;**289**:102362. DOI: 10.1016/j.cis.2021.102362

[61] Lee WJ et al. Fouling mitigation in forward osmosis and membrane distillation for desalination. Desalination. 2020;**480**. DOI: 10.1016/j.desal.2020.114338

[62] Deshmukh A et al. Membrane distillation at the water-energy nexus: Limits, opportunities, and challenges. Energy & Environmental Science. 2018;**11**(5):1177-1196. DOI: 10.1039/c8ee00291f

Chapter 2

Thermal Desalination Systems: From Traditionality to Modernity and Development

Fadl A. Essa

Abstract

As well known, the basic birthrights of human are the clean air, clean water, healthy food, and green energy. So, clean water is the second important requested need of all living organisms on Earth. To know the importance of water to our human bodies, a deficiency of just 2% in our body's water supply indicates dehydration. Nowadays, all countries suffer from the problem of freshwater shortage. Despite the importance of clean water for our lives, only 0.01% is available as surface water such as the rivers, lakes, and swamps. These frightening facts have made it a national and humanitarian duty for scientists to research how to overcome the water problem and how to provide alternative sources of safe drinking water using renewable energies. Desalination is the most famous and operative technique used to overcome this problem. In this chapter, the different desalination techniques are reviewed and reported. Also, the solar distillation processes are mentioned with an extended review on the solar distillers. Besides, the application of artificial intelligence in improving the performance of desalination systems is reported. The main conclusions are stated at the end of this chapter.

Keywords: desalination, thermal desalination processes, stage flash, reverse osmosis, solar distiller

1. Introduction

Any human being needs the water as the second most important fluid after air to be able to live on the Earth. Actually, around two-thirds of the Earth is covered by water (~ 71%), but more than 7% of this water cannot be used because they are in the form of ocean, ice caps, glaciers, ground, and aquifers [1]. Therefore, the freshwater that is available to use by the people is only around 1% all over the globe. So, the need for desalinated freshwater arises day by day [2]. In addition, the applications of freshwater such as cooking, drinking, and farming make it in a difficult situation. As a result, providing safe drinking water is a major challenge all over the world [3, 4]. The predicted freshwater shortage problem in 2030 is illustrated in **Figure 1**.

The simplest well-known cycle for the water is the hydrological cycle, which had sequentially the steps of water surface evaporation, condensation, cloud creation, runoff, and rain. So, the water controls the ecosystem of any society [6, 7]. This is because the potable water shortage is a problem of both the remote and urban communities [8].

Global water scarcity -2030

62% * of world population

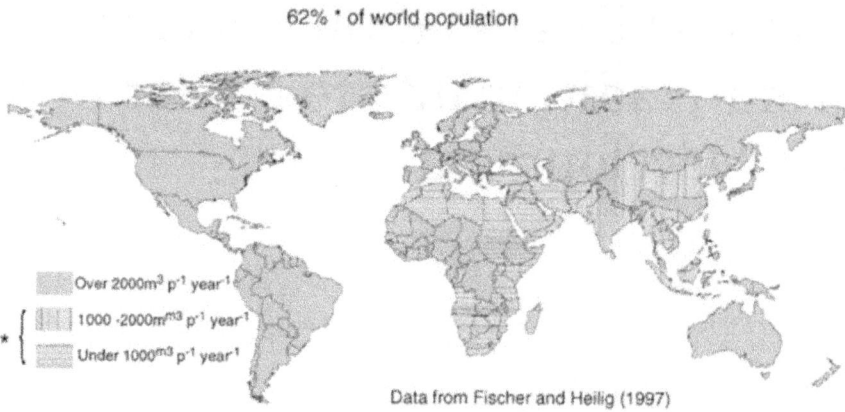

Over 2000m³ p⁻¹ year⁻¹

1000 -2000m³ p⁻¹ year⁻¹

Under 1000m³ p⁻¹ year⁻¹

Data from Fischer and Heilig (1997)

Figure 1.
Estimated global water scarcity in 2030 [5].

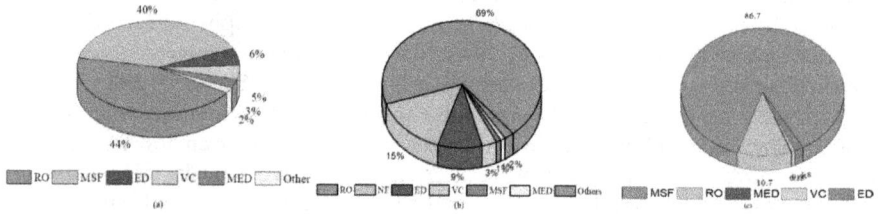

Figure 2.
Capacity of distillation processes for (a) the globe, (b) USA, an (c) Middle East countries in 2002 [16].

The thermal desalination processes follow the same principles of the natural hydrological cycle, but they use huge energy amounts. Distillation is accomplished by introducing saltwater into the process, which produces two output streams: one of freshwater and the other of brine water. Distillation of seawater produces freshwater [9–13].

Distillation becomes a key source of freshwater for most of the world's regions. Because of the ample water, the distillation procedure is mostly considered in coastal places. The most significant feature of this procedure is that it is completely safe for everyone involved—in other words, it has no negative consequences for the environment [14]. According to a survey conducted in the preceding decade, roughly 75 million people around the world rely on distillation for their everyday requirements. Distillation is the only source of freshwater for many countries. Saudi Arabia, the United States, the United Arab Emirates, Spain, and Kuwait are the top five countries in terms of desalination plant capacity, with 17.4, 16.2, 14.7, 6.4, and 5.8%, respectively [15]. The capacity of distillation processes for various countries is depicted in **Figure 2**; for (a) the globe, (b) USA, and (c) Middle East countries in 2002.

Till 2015, the total capacity of desalinated freshwater is 86.55 million cubic meter a day, which were obtained from around 18,000 plants for desalination around the globe. The Middle East and North Africa account for almost 44% of the above total capacity. Over the last 20 years, 80% of the energy utilized for freshwater producing has been lowered because of contemporary advancements in distillation technologies [16].

2. Desalination techniques

Desalination of brackish/saline water can be done in a variety of ways. Among the distillation methods, multistage flash desalination (MSFD), multi-effect distillation (MED), and reverse osmosis (RO) are considered the most commercial and economic viable technologies for distillation. These techniques are the most effective, and they will continue in the future [17]. Desalination can be done in several ways, as listed below.

2.1 Multistage flash distillation (MSFD)

MSFD is based on flashing of evaporation. Evaporation of saltwater occurs as a result of a decrease in pressure rather than a rise in temperature. Broadly speaking, regenerative heating is used to achieve a maximal production and best MSFD's economics. The seawater flashing in the flashing chamber regenerates and transfers its thermal energy to the saltwater passing *via* the flashing action. Because this is a regenerative heating system, it must be completed in stages. Then, it would be better if we increased the incoming saline water temperature [18]. The condensation heat (released when condensating the water vapor) is responsible for the progressive raise in saline water temperature. The heat input, heat recovery, and heat rejection are the most important factors for MSFD unit [19].

Multistage evaporators, with roughly 19–28 stages, are utilized in recent MSFD plants [20]. The MSFD operating temperatures are between 90 and 120°C. The more the operating temperature of unit, the higher the efficiency of the system. Also, the pressure should be controlled under the water saturation temperature.

In the unit (see **Figure 3**), various devices or accessories such as demisters, decarbonators, and vacuum deaerators are employed for various reasons. Demisters are used to prevent the carryover of saline water into condensed distillate. Removing the dissolved gases in brine is the function of decarbonator and vacuum deaerator [22].

2.2 Multiple-effect distillation (MED)

MED method is the earliest of all distillation techniques. Thermodynamically, it is the most constitutional technique among all other distillation methods. The term "effect" of the MED name refers to how many evaporators' series. Then, the main fundamental of MED is the decrease in ambient pressure through various effects. In this process, no additional heat is required after the first effect. This is due to that

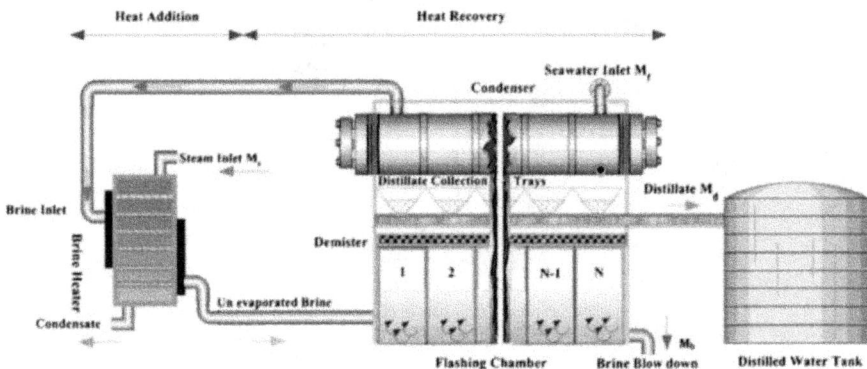

Figure 3.
Illustration of processes of MSFD [21].

MED automatically makes the saline water to be boiled many times through the corresponding pressure.

In this process, the saline water temperature is increased to its boiling value at the first effect through the heat exchanger tubes. The entering saline water is sprayed over the evaporator surface for better vapor generation. The created vapor is condensed on the other side of tubes. On the second hand, the condensation process is utilized on the same time to heat up the saline feedwater. **Figure 4** illustrates the desalination with solar energy.

MED economy relies on its effects number regarding its chained processes [8]. So, the first effect has some generated vapor from the seawater entering the effect. After that, the remaining water is fed to the next effect. The amid tubes are warmed up using the first-effect created vapor. Also, that vapor is condensated to produce potable water. Besides, the generated vapor heats up the remaining saline water of the next effect. This action is repeated from one effect to another under low pressure and temperature till the end of process with around 4–21 effects [23].

2.3 Vapor compression distillation (VCD)

As it is known, raising the pressure of steam vapor leads to increase its temperature, and hence, the heat released from this vapor is increased. This concept is utilized in the VCD method, where the vapor-increased heat is utilized to evaporate the incoming

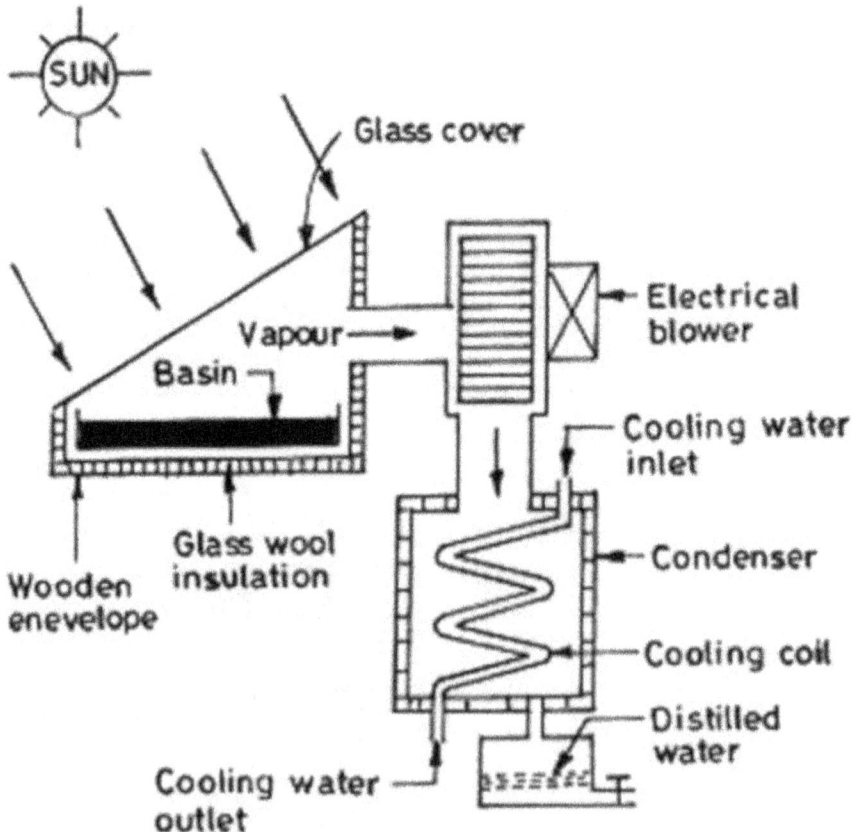

Figure 4.
Desalination by solar energy [2].

saline water. So, the main key of VCD is reducing the medium pressure, which leads to decrease the boiling temperature of water [24]. This process can be achieved by either steam jet technique (thermo-compressor technique) or mechanical compressor device (electrically driven technique), and these techniques are utilized to condense vapor content into distillate and use the corresponding latent heat as a heat source for evaporating the incoming saline water.

In the technique of thermo-compressor, a venturi orifice is used in the stream of steam jet to create low pressure. After that, the vapor content is compressed *via* the steam jet with the help of the venturi orifice device. Therefore, the generated vapor content is condensated over the tubes' surfaces with releasing heat to vaporize the incoming salt water.

In addition, there is another kind of VCD, which depends on lowering the temperature inside the VCD, and hence, this kind needs only power to run. This technique of VCD is useful for small desalination units due to its simple construction, better efficiency, and process reliability [25]. As a result, VCD is applicable in resorts, industries, and drilling sites.

2.4 Reverse osmosis (RO)

It is known that the water flows naturally from the freshwater direction to saltwater direction. RO depends mainly on a critical parameter called osmotic pressure, which should force the saltwater to flow in the opposite direction of natural flow. As a result, we need an external source of pressure to overcome the osmotic pressure. That external pressure must be more than the osmotic pressure. Hence, the name of this method (reverse osmosis) refers to the process meaning reversing the direction of normal water flow through the membrane. By applying this process, the salts in the saline water are left behind and not allowed to cross the membranes [26]. This process produces potable water (permeate) and brine water (concentrate) as illustrated in **Figure 5**. Also, **Figure 5** illustrates the main components of RO unit such as the pretreatment, membranes, high-pressure pumps, and post-treatment.

The pretreatment process is important for eliminating the undesirable materials that damage the membrane [27]. It relies on the membrane kind and configuration, properties of feedwater, recovery ratio, and required permeate quantity and quality. According to those parameters, the pretreatment process has different techniques to be applied such as chlorination, coagulation, acid addition, micron cartridge filtration, multimedia filtration, and dechlorination. Moreover, to overcome the osmotic pressure, high-pressure pumping system such as the centrifugal pump is utilized. Additionally, the different membrane configurations such as the spiral wound and hollow fine fiber (HFF) strongly affect the performance of the RO unit [28]. Besides, adjusting the pH and adding H_2S and CO_2 are performed in the post-treatment process [29].

2.5 Freezing

Another type of desalination processes is nominated freezing, which is simple to conduct and operate. This process depends mainly on the fact that the dissolved salts are removed while forming the ice crystals from the saline water. First, we clean and wash the saline water mixture to leave the salt in the left water behind before freezing the whole water. Then, we can get the freshwater *via* melting the ice crystals. As a result, freezing has the processes of saline water cooling, partial creation of ice, separation of ice from saline water, ice melting (obtaining freshwater), and finally refrigeration and heat rejection [30].

Permeate
TDS: 166 – 365 mg/L

RO Feed
TDS: 40,070 mg/L

Recovery: 45%

Concentrate
TDS: 72,500 – 72,700 mg/L

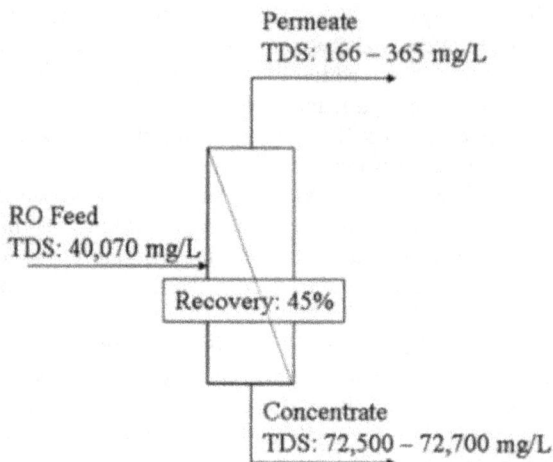

Figure 5.
Schematic of RO desalination technique [16].

Freezing units have the merits of low power consumed, few corrosions, and eliminated scaling factors. But it has the demerits of handling water and ice mixtures because it is mechanically hard to treat. Unfortunately, this method still needs numerous improvements to be applied on the commercial level. Limited freezing stations exist all over the globe such as that was built in Saudi Arabia [31].

2.6 Solar desalination

Here, the sun energy is the driving force for such solar desalination systems, which actually are alike the natural hydrological cycle. The natural hydrological cycle takes every day in the form of heating the seawater using the solar radiation to produce vapor, and this vapor content is condensated due to the low temperature in the heights. An application on the solar distillation systems is the greenhouse solar distiller [25, 32].

2.7 Potabilization

This process is almost linked to MSFD systems. This is because MSFD produces distillate with small impurities amounts, and then, the potabilization is utilized to eliminate these impurities [33]. The potabilization process can be conducted *via* two different methods: injecting CO_2 and hydrated lime [34] and carbonated water is passed through limestone bed filters [35]. Potabilization has four main processes: carbonation, liming, chlorination, and aeration. The main functions of liming and carbonation are increasing the hardness, alkalinity, mineral content, and pH of the targeted water. Also, chlorination (performed *via* using chlorine gas or calcium hypochlorite) aims at avoiding the infected water [36], while aeration aims at replacing the oxygen inside the water to enhance its taste.

3. Desalination economics

The economics of any system determine its success. In desalination stations, the parameters of fixed cost, running costs, station location, maintenance cost, and energy consumption are the controlling factors of the station economics. There are two opposite directions for determining the economics of the desalination systems. First is the improvements in desalination systems, which leads to reduce the cost of the system as a whole. Second is the pollution and contamination of water, which raises the cost of the desalination system.

The economics and technical properties of the desalination method with the targeted quantity of freshwater are the factors based on which we select the distillation technique. The technical properties include the energy-driving source, consumed energy, freshwater specifications, land space of unit, station reliability, operational aspects, plant size, and the maintenance of spare parts, while the economic parameters include the fixed costs of the station, operating costs, interest rate, life cycle of station, and maintenance costs ...etc. [37]. The cost of distilled freshwater is described by $/m^3. The cost of desalinated water is determined using the following equation.

$$Cost\ of\ desalinated\ water = \frac{All\ spent\ and\ estimated\ costs\ through\ station\ lifetime}{Total\ produced\ water\ quantity} \quad (1)$$

4. Solar distillers

As explained before, solar distillation is one of the desalination techniques. Solar distiller is one of the solar distillation devices' family. It is a simple in construction, cheap, and easy to maintain device [38]. But it has the demerit of few freshwater production. As a result, the scientists do their best to improve the yield of solar distillers [39–45]. The solar distiller has the parts of glazing cover, basin, collecting trough, and some instruments as shown in **Figures 6** and 7. The basin is fed by saline water to be heated and vaporized inside the distiller. Then, the vapor is condensed on the inner glazing surface. After that, the condensated droplets are taken out using a hose into the calibrated flasks. The surfaces of the solar distiller are painted with black for maximizing the absorbed solar radiations. Also, the device body is insulated by saw dust or fiberglass for avoiding the thermal losses. Moreover, the measuring instruments are utilized to be able to evaluate the solar distiller performance.

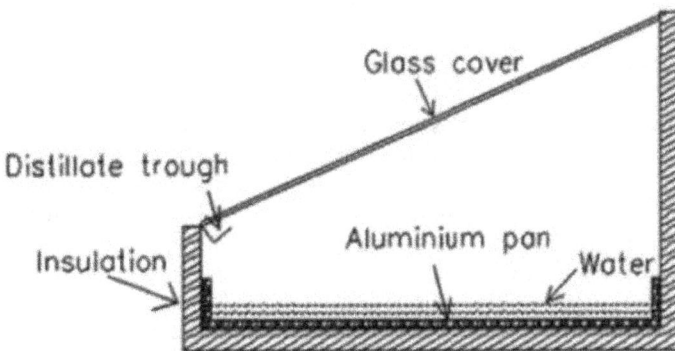

Figure 6.
Single-basin distiller [46].

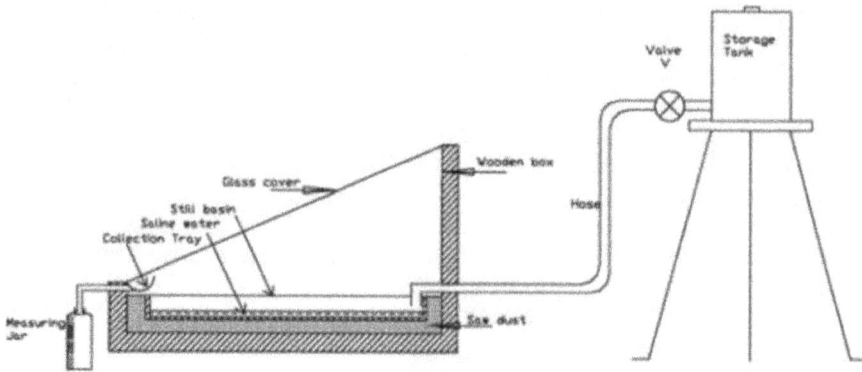

Figure 7.
Solar still with simple basin [40].

5. Methods of improving the solar distiller performance

One of the biggest problems of the twenty-first century is the global freshwater scarcity, which has numerous side effects on the mankind [47, 48]. The widespread of the water problem on a global level has made more impacts on the lives of people who live far from urbanization and remote areas, and more and more on the lives of the poor who do not have the costs of using high technology to produce water. As a result, the science developed various high and low water desalination technologies to be used at the level of industrial and commercial production and at the level of individuals and families [49]. The water desalination technologies can be categorized by membranes or thermal processes. Nevertheless, the high technologies demand building complex and large central installations, which causes them to be infeasible for developing regions such as distributed, poor, and remote areas. In addition, the rural, arid, and remote areas need desalination methods with no or minimum maintenance, supervising, and operating requirements [41, 45]. Consequently, the solar-powered desalination units such as solar distillers meet all these conditions, which make them as an efficient candidate to provide drinkable water in these regions [42, 43].

Nevertheless, the solar distillers have low output productivity (1.5–2.5 L/m^2 day) and low thermal efficiency ($\approx$ 30%), which are the main bottlenecks of this distillation technique [50–52]. As a result, numerous investigations focused on improving the performance of solar distillers. Consequently, the solar stills such as stepped type [53–55] (the absorber of the basin takes the steps shape), disk type [56] (the main effective absorber is a rotating disk), vertical type [57] (the distiller horizontal width is very small compared to its vertical height), tubular type [58–60] (the outer shape of the distiller is tubular/cylindrical), drum type [61, 62] (the absorber is a rotating drum inside the basin still), PV/T active type [63, 64] (distiller powered by PV panels), finned type [65] (the absorber of the distiller is a collection of fins), trays type [66–69] (the main effective absorber has trays to enlarge the surface area), inclined type [70] (the absorber is inclined), wick type [10, 71, 72] (the wick material is spread over the absorber of distiller), corrugated type [73, 74] (the absorber of the distiller is a collection of corrugated shapes), spherical type [75] (the distiller takes the shape of sphere), double-effect type [76–78] (the distiller has two stages of water basins), multistage type [79] (the distiller has multistages of water basins), inverted absorber type [80] (the distiller has inverted absorber inside it), hemispherical type [81] (the distiller takes the shape of sphere), convex type [82, 83] (the absorber has a convex shape), and pyramid type [84–86] (the distiller takes the form of pyramid) are found in the literature. In addition, numerous modifications were performed to improve the distiller performance. These modifications included the use of condensation unit [87–89], nanofluids [90, 91], heat exchanger [92], floating aluminum sheet [93], desiccant [94], solar ponds [95], glass cooling [44], volcanic rocks [96], wick materials [71], rotating parts [97–99], coated absorbers [100, 101], phase change materials (PCM) [102–104], fins [105], half barrel and corrugated absorbers [106], solar collectors [107], sun-tracking systems [108], multiple-effect basins [109], reflectors [110], vapor extraction [111–113], and reusing latent heat of evaporation [114].

6. Solar distiller types

The solar distillers can be classified into single-effect and multi-effect distillers (according to the number of effects of distiller) with a subcategory of active and passive distillers (according to the vaporization heat source) inside every classification [2]. In the passive distillers, the vaporization occurs directly without external sources of heat, while using external heat sources such as collectors and concentrators are used in the active solar distillers.

6.1 Single-effect distillers

This is the traditional solar distiller (or conventional distiller) without any modifications [115]. Also, it is used as a reference for the other modified distillers' performances. Here, in this type of distillers, there is just one glass cover over the basin water. Also, the thermal losses in this type of distiller are large due to the single glass cover, which hence decreases its performance. Therefore, the efficacy of this distiller is low around 30–40% [2]. As a result, numerous experimental and empirical investigations were performed to augment the distiller performance. This type of distillers (single-effect solar distillers) is categorized into active and passive distillers.

6.1.1 Active solar distiller

The word "active" means that the solar distiller is integrated with somewhat external source of heat such as the solar concentrators and collectors [2]. Then, the active distillers include the following items:

1. Regenerative distillers

2. Air bubbled distillers

3. Waste heat recovery distillers

4. Distiller with heat exchangers

5. Distiller integrated with concentrators

6. Distiller incorporated with hybrid units

7. Distiller incorporated with heaters

Figure 8 shows the active distiller integrated with evacuated collector.

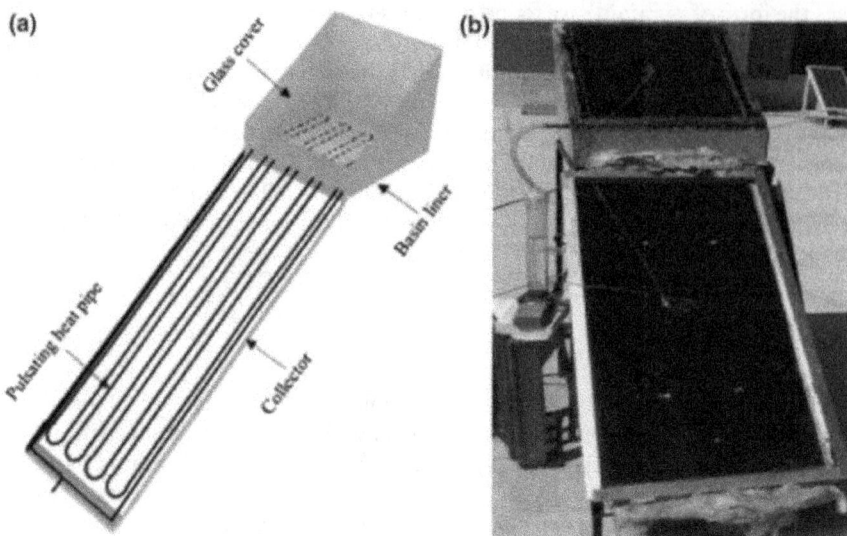

Figure 8.
Active distiller integrated with evacuated collector [116].

6.1.2 Passive solar distiller

Here, the heat source for basin water vaporization is only from inside of distiller [2]. Then, the passive solar distillers include the following kinds.

1. Basin distiller (conventional distiller).

2. Wick distiller.

3. Weir-type distiller.

4. Spherical distiller.

5. Tubular distiller.

6. Pyramid distiller.

7. Diffusion distiller.

8. Greenhouse combination distiller.

9. Stepped distiller.

For example, **Figure 9** obtains a passive distiller incorporated with condenser, and **Figure 10** reveals a passive distiller incorporated with internal and external mirrors.

6.2 Multi-effect distillers

Multi-effect solar distillers are different in design from the single-basin distillers. Also, the modified design of multi-effect distillers helped enhancing

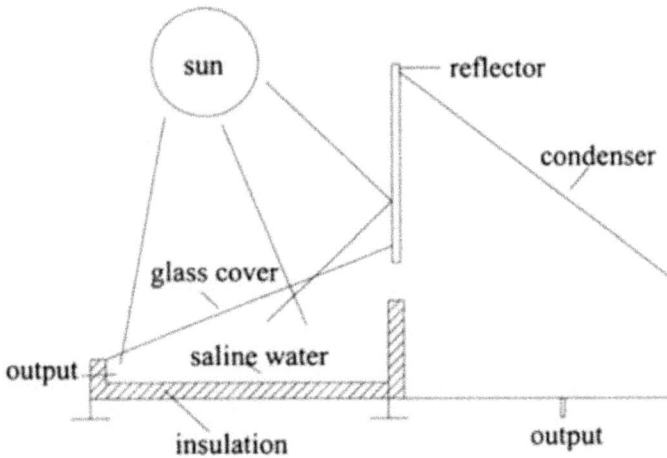

Figure 9.
Passive distiller integrated to condenser [117].

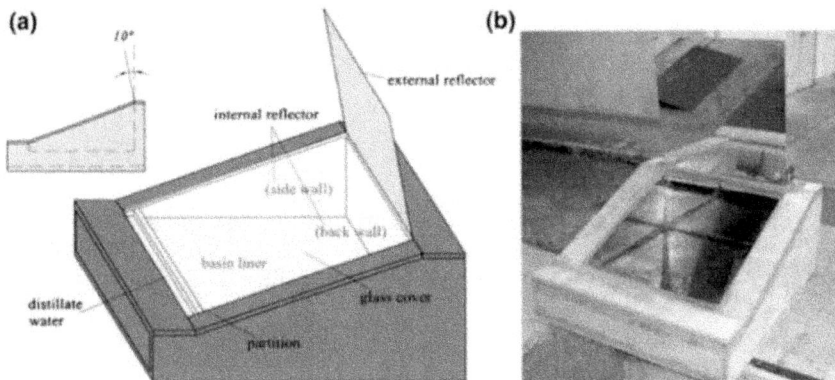

Figure 10.
Passive distiller integrated with mirrors: Schematic and photo diagrams [118].

the performance of the distillers. Here, in these types of distillers, the condensation latent heat is utilized as a recovery source of heat to obtain more vaporization through the effect of the distillers, and hence, the freshwater production is augmented [119]. Now, multi-effect solar distillers are categorized into two main sections: active and passive distillers as the following.

6.2.1 Active distiller

It is the same concept of the single-effect distiller. The word "active" means that the solar distiller is integrated with somewhat external source of heat such as the solar concentrators and collectors. Various distiller kinds can be found in the literature under the category of active distiller-based multi-effect distillers.

1. Multistage evacuated active distiller.

2. Multi-basin inverted absorber active distiller.

3. Waste heat recovery active distiller.

4. Multi-effect condensation–evaporation desalination unit.

5. Distiller integrated with collectors such as flat plate and tube collector.

A multi-effect active distiller with collector is illustrated in **Figure 11**. Also, **Figure 12** shows a condensation-evaporation active distiller.

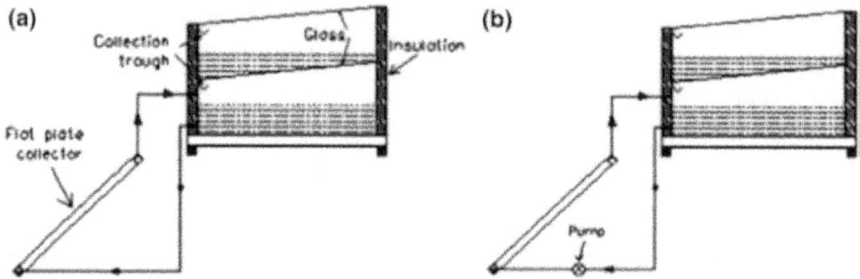

Figure 11.
Double-effect single-slope active still: a coupled with solar collector in thermosiphon mode and b coupled with solar collector in forced circulation mode [120].

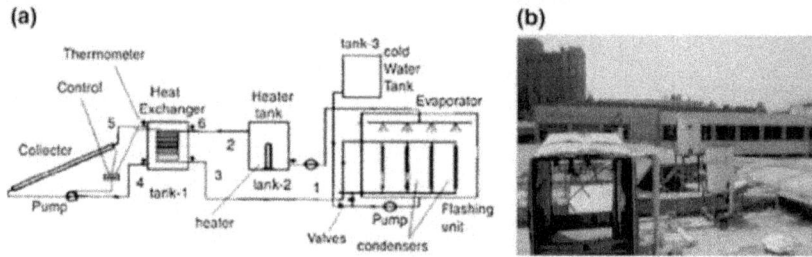

Figure 12.
Condensation–evaporation active distiller [121].

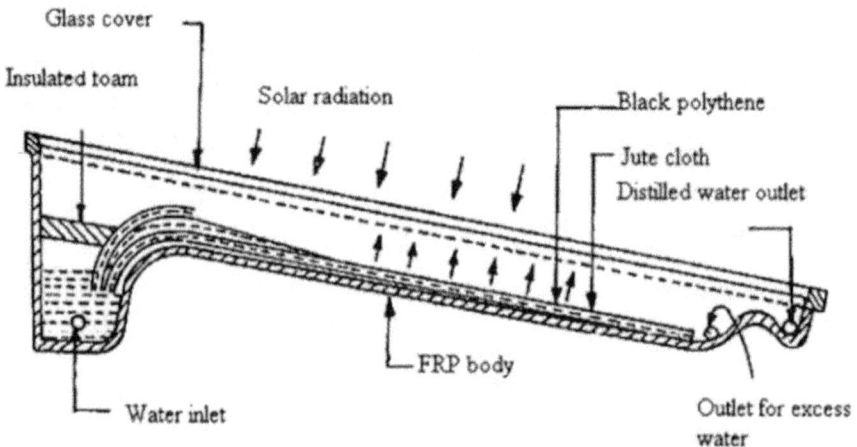

Figure 13.
Wick-passive distiller [122].

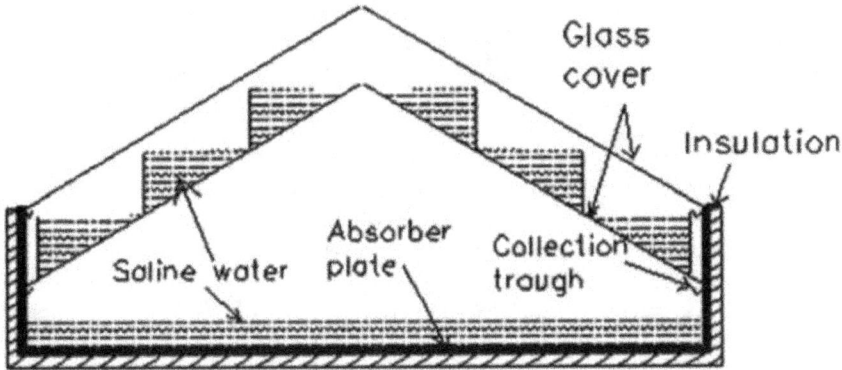

Figure 14.
Double-effect double-basin passive distiller [46].

6.2.2 Passive distiller

Here, the passive multi-effect distillers include the following designs of distillers.

1. Wick distillers.

2. Conventional distillers.

3. Weir-type distillers.

4. Diffusion distillers.

Figure 13 illustrates a passive wick distiller, and **Figure 14** shows a double-basin double-effect passive distiller.

7. ANN as a prediction method for the performance of desalination systems

Artificial intelligence (AI) is quickly progressing every day. Computers can easily perform the difficult tasks for the humans to do. AI is first contrived in 1950. AI passed with steps and stages; some failed and others succeeded. Deep learning with powerful computers is the main keys of the success of AI. With time, AI can solve numerous industrial problems including the image processing, recognition of speech, language processing, optimization, prediction, robotic automation, categorization, self-driving cars ...etc. AI is of numerous branches, and we will focus, in this chapter, on the neural networks (NNs). NNs help the computers to teach themselves from observing the data to predict the system performances. Preparing data for neural network processing is typically the most difficult and time-consuming task you will encounter when working with neural networks. In addition to the enormous volume of data that could easily reach millions and even billions of records, the main difficulty is in preparing the data in the correct format for the task in question.

7.1 Application of ANN on desalination systems

Contemporary experience indicates that artificial neural networks (ANN's) may be specifically appropriate to offer tools to assist desalination plant operators in day-to-day operations. Prediction of the productivity of the desalination units is an essential consequence to evaluate its potentiality to provide potable water with involving in conducting more experiments. AI-based methods are stated as robust tools to obtain the correlation between the process parameters and responses [123–127]. This chapter suggests that ANNs are particularly appropriate as the basis for the development of tools to aid in the various phases of operating a desalination plant. In solar energy applications, different kinds of ANNs predicted and optimized the performance of collectors, cells, distillers, etc. [128]. The most application of ANNs is solving the issues of designing and procedures in electric power systems. Due to the similarities between power plants and desalination plants, the methods of ANNs can be used to predict the performance of desalination plants [129]. The merits of applying ANN included that it handles large amounts of datasets, can discover complex nonlinear relationships among all parameters (dependent and independent), and can find all interactions among all tested parameters.

Zarei and Behyad [130] introduced an ANN to test the effective parameters of the greenhouse on water productivity such as width, length, height of evaporator, and roof transparency. The method obtained an acceptable agreement with the experimental results. Tayyebi and Alishiri [131] used a nonlinear inverse model control strategy based on neural network for predicting the performance of MSF desalination plant. The proposed NNs consisted of three layers identified from input–output data and trained with a descent gradient algorithm. The set point tracking performance of the proposed method was studied when the disturbance is present in MSF unit. Three controllers were provided for checking the saline temperature, last stage level, and salinity. The results obtained that a neural network inverse model control strategy is robust and highly promising to be implemented in such nonlinear systems. An ANN model based on field data was built to investigate the vacuum membrane distillation (VMD) performance at various factors [132]. The introduced model could accurately predict the unseen data of VMD. The correlation coefficient of the overall agreement between ANN predictions and experimental data was found to be more than 0.994. Aish et al. [133] utilized ANN to forecast reverse osmosis performance in Gaza Strip through predicting the next week values of total dissolved solids (TDS) and permeate flow rate of the product water. Multilayer perceptron (MLP) and radial basis function (RBF) neural networks were trained and developed with reference to feedwater parameters including the pressure, pH, and conductivity to predict permeate next week values of flow rate. MLP and RBF neural networks were used for predicting the next week TDS concentrations. Both networks are trained and developed with reference to product water quality variables including the water temperature, pH, conductivity, and pressure. The prediction results showed that both types of neural networks are highly satisfactory for predicting TDS level in the product water quality and satisfactory for predicting permeate flow rate.

Santos et al. [134] simulated the distiller yield based on local weather data such as the solar irradiance, air temperature, glass temperature, and air speed and direction by ANNs. Essa et al. [51] introduced an enhanced ANNs incorporated to Harris Hawks optimizer to anticipate the distiller yield. The optimizer was utilized to get the optimal structure and network factors. Nevertheless, traditional artificial neural networks suffer from trapping in local optima problem during learning process. To overcome this problem, RVFL is proposed as it determines the output weights without involving in time-consuming learning process. Besides, RVFL has a direct connection among the input and output nodes, which is of a substantial impact on the network performance. This aids in avoiding the overfitting issue occurring in other ANNs. Using these merits, the conventional RVFL networks has some restrictions like that there are various conformations can be utilized, and this can affect the quality of water productivity final prediction. Then, Ensemble Random Vector Functional Link Networks (EnsRVFL) [135], which relies on RVFL as a basic modeling, were utilized to eliminate these restrictions of RVFL. A random vector functional link (RVFL) network integrated with artificial ecosystem-based optimization (AEO) algorithm was introduced to predict the seawater greenhouse (SWGH) performance [52]. Power consumption and yield of SWGH were predicted *via* RVFL-AEO modeling. Besides, the statistical analyses with various statistical tools were conducted to find out the neural network efficacy. The statistical tools revealed a complete match between the field and modeling data. The RVFL-AEO performance was compared with that of conventional RVFL. RVFL-AEO obtained an improved performance as compared by RVFL, which indicates the role of AEO in obtaining the optimal RVFL factors that improved the model accuracy.

8. Conclusions

Based on the above explained sections, the following points can be concluded. The most economic and efficient method to get freshwater is desalination. Demisters are the main component of MSFD technique to prevent mixing the distillate by brine. The thermal performance of MED is better than that of MSFD. For example, GOR is 10 and 15 for MSFD and MED, respectively. RO is considered as the most commercial, economic, and efficient techniques of distillation. The economics and technical properties of the desalination method with the targeted quantity of freshwater are the factors based on which we select the distillation technique. Solar distillers are the simplest and easiest way to get potable water, but they suffer from the few yields. The solar distillers can be classified into single-effect and multi-effect distillers (according to the number of effects of distiller) with a subcategory of active and passive distillers (according to the vaporization heat source) inside every classification.

Author details

Fadl A. Essa
Faculty of Engineering, Mechanical Engineering Department, Kafrelsheikh
University, Kafrelsheikh, Egypt

*Address all correspondence to: fadlessa@eng.kfs.edu.eg

IntechOpen

References

[1] Tiwari GN, Sahota L. Advanced Solar-Distillation Systems: Basic Principles, Thermal Modeling, and Its Application. Singapore: Springer; 2017

[2] Kumar PV, et al. Solar stills system design: A review. Renewable and Sustainable Energy Reviews. 2015;**51**:153-181

[3] Tiwari G, Singh H, Tripathi R. Present status of solar distillation. Solar Energy. 2003;**75**(5):367-373

[4] McCutcheon JR, McGinnis RL, Elimelech M. A novel ammonia—carbon dioxide forward (direct) osmosis desalination process. Desalination. 2005;**174**(1):1-11

[5] Wallace JS. Increasing agricultural water use efficiency to meet future food production. Agriculture, Ecosystems & Environment. 2000;**82**(1):105-119

[6] Pahl-Wostl C. Transitions towards adaptive management of water facing climate and global change. Water Resources Management. 2007;**21**(1): 49-62

[7] Kummu M, et al. Is physical water scarcity a new phenomenon? Global assessment of water shortage over the last two millennia. Environmental Research Letters. 2010;**5**(3):034006

[8] Khawaji AD, Kutubkhanah IK, Wie J-M. Advances in seawater desalination technologies. Desalination. 2008;**221**(1-3):47-69

[9] Qiblawey HM, Banat F. Solar thermal desalination technologies. Desalination. 2008;**220**(1):633-644

[10] Abdullah A, et al. Performance evaluation of a humidification–dehumidification unit integrated with wick solar stills under different operating conditions. Desalination. 2018;**441**:52-61

[11] Sharshir SW, et al. Performance enhancement of wick solar still using rejected water from humidification-dehumidification unit and film cooling. Applied Thermal Engineering. 2016; **108**:1268-1278

[12] Essa FA, et al. On the different packing materials of humidification–dehumidification thermal desalination techniques – A review. Journal of Cleaner Production. 2020;**277**:123468

[13] Abdullah A, et al. An augmented productivity of solar distillers integrated to HDH unit: Experimental implementation. Applied Thermal Engineering. 2020;**167**:114723

[14] Lattemann S, Höpner T. Environmental impact and impact assessment of seawater desalination. Desalination. 2008;**220**(1):1-15

[15] Templer KJ, Tay C, Chandrasekar NA. Motivational cultural intelligence, realistic job preview, realistic living conditions preview, and cross-cultural adjustment. Group & Organization Management. 2006;**31**(1):154-173

[16] Greenlee LF, et al. Reverse osmosis desalination: water sources, technology, and today's challenges. Water Research. 2009;**43**(9):2317-2348

[17] Saito K, et al. Power generation with salinity gradient by pressure retarded osmosis using concentrated brine from SWRO system and treated sewage as pure water. Desalination and Water Treatment. 2012;**41**(1-3):114-121

[18] Kamaluddin BA, Khan S, Ahmed BM. Selection of optimally matched cogeneration plants. Desalination. 1993;**93**(1-3):311-321

[19] Consonni S. Optimization of cogeneration systems operation. Part A: Prime movers modelization, in:

Proceedings of the American Society of Mechanical Engineers Cogen-Turbo international symposium on turbomachinery, combined-cycle technologies and cogeneration. France: Nice; 1989

[20] Sommariva C, Syambabu V. Increase in water production in UAE. Desalination. 2001;**138**(1-3):173-179

[21] Baig H, Antar MA, Zubair SM. Performance evaluation of a once-through multi-stage flash distillation system: Impact of brine heater fouling. Energy Conversion and Management. 2011;**52**(2):1414-1425

[22] Finan M, Harris A, Smith S. The field assessment of a high temperature scale control additive and its effect on plant corrosion. Desalination. 1977; **20**(1-3):193-201

[23] Michels T. Recent achievements of low temperature multiple effect desalination in the western areas of Abu Dhabi. UAE. Desalination. 1993; **93**(1-3):111-118

[24] Council NR. Committee on Advancing Desalination Technology. In: Desalination: A National Perspective. The National Academies Press, National Research Council; Division on Earth and Life Studies; Water Science and Technology Board; Committee on Advancing Desalination Technology; 2008

[25] Buros O. The ABCs of Desalting. MA: International Desalination Association Topsfield; 2000

[26] Baig MB, Al Kutbi AA. Design features of a 20 migd SWRO desalination plant, Al Jubail, Saudi Arabia. Desalination. 1998; **118**(1-3):5-12

[27] Al-Sheikh, A.H.H., Seawater reverse osmosis pretreatment with an emphasis on the Jeddah Plant operation

experience. Desalination 1997;**110**(1-2): 183-192

[28] Whitesides GM. Whitesides' group: Writing a paper. Advanced Materials. 2004;**16**(15):1375-1377

[29] Avlonitis S, Kouroumbas K, Vlachakis N. Energy consumption and membrane replacement cost for seawater RO desalination plants. Desalination. 2003;**157**(1):151-158

[30] Buros O. Conjunctive use of desalination and wastewater reclamation in water resource planning. Desalination. 1976;**19**(1-3):587-594

[31] Hanahan D, Weinberg RA. Hallmarks of cancer: The next generation. Cell. 2011;**144**(5):646-674

[32] Chang F, et al. Bigtable: A distributed storage system for structured data. ACM Transactions on Computer Systems (TOCS). 2008; **26**(2):1-26

[33] Withers A. Options for recarbonation, remineralisation and disinfection for desalination plants. Desalination. 2005;**179**(1-3):11-24

[34] Khawaji AD, Wie J-M. Potabilization of desalinated water at Madinat Yanbu Al-Sinaiyah. Desalination. 1994;**98**(1-3):135-146

[35] Al-Rqobah H. A recarbonation process for treatment of distilled water produced by MSF plants in Kuwait. Desalination. 1989;**73**:295-312

[36] Ayyash Y, et al. Performance of reverse osmosis membrane in Jeddah Phase I plant. Desalination. 1994; **96**(1-3):215-224

[37] Wade NM. Technical and economic evaluation of distillation and reverse osmosis desalination processes. Desalination. 1993;**93**(1-3):343-363

[38] Nafey A, et al. Solar still productivity enhancement. Energy Conversion and Management. 2001; **42**(11):1401-1408

[39] Al-Hayeka I, Badran OO. The effect of using different designs of solar stills on water distillation. Desalination. 2004;**169**(2):121-127

[40] Velmurugan V, et al. Single basin solar still with fin for enhancing productivity. Energy Conversion and Management. 2008;**49**(10):2602-2608

[41] Abdullah A, Essa F, Omara Z. Effect of different wick materials on solar still performance–a review. International Journal of Ambient Energy. 2019;**42**(9):1-28

[42] Elsheikh A, et al. Applications of nanofluids in solar energy: A review of recent advances. Renewable and Sustainable Energy Reviews. 2018; 82:3483-3502

[43] Kabeel AE, et al. Solar still with condenser – A detailed review. Renewable and Sustainable Energy Reviews. 2016;**59**:839-857

[44] Omara Z, et al. The cooling techniques of the solar stills' glass covers–a review. Renewable and Sustainable Energy Reviews. 2017; 78:176-193

[45] Panchal H, et al. Enhancement of the yield of solar still with the use of solar pond: A review. Heat Transfer. 2020;**n/a(n/a)**

[46] Rajaseenivasan T, et al. A review of different methods to enhance the productivity of the multi-effect solar still. Renewable and Sustainable Energy Reviews. 2013;**17**:248-259

[47] Mekonnen MM, Hoekstra AY. Four billion people facing severe water scarcity. Science Advances. 2016; 2(2):e1500323

[48] Shannon M, Bohn P, Elimelech M, et al. Science and technology for water purification in the coming decades. Nature. 2008;**452**:301-310. https://doi.org/10.1038/nature06599

[49] Elimelech M, Phillip WA. The future of seawater desalination: Energy, technology, and the environment. Science. 2011;**333**(6043):712-717

[50] Abd Elaziz M, Essa F, Elsheikh AH. Utilization of ensemble random vector functional link network for freshwater prediction of active solar stills with nanoparticles. Sustainable Energy Technologies and Assessments. 2021; 47:101405

[51] Essa FA, Abd Elaziz M, Elsheikh AH. An enhanced productivity prediction model of active solar still using artificial neural network and Harris Hawks optimizer. Applied Thermal Engineering. 2020;**170**:115020

[52] Essa FA, Abd Elaziz M, Elsheikh AH. Prediction of power consumption and water productivity of seawater greenhouse system using random vector functional link network integrated with artificial ecosystem-based optimization. Process Safety and Environmental Protection. 2020; **144**:322-329

[53] Essa F, et al. Wall-suspended trays inside stepped distiller with Al_2O_3/paraffin wax mixture and vapor suction: Experimental implementation. Journal of Energy Storage. 2020;**32**:102008

[54] Essa FA, Omara Z, Abdullah A, et al. Augmenting the productivity of stepped distiller by corrugated and curved liners, CuO/paraffin wax, wick, and vapor suctioning. Environ Sci Pollut Res; 2021;**28**:56955-56965. https://doi.org/10.1007/s11356-021-14669-w

[55] Shanmugan S, et al. Performance of stepped bar plate-coated nanolayer of a box solar cooker control based on

adaptive tree traversal energy and OSELM. In: Machine Vision Inspection Systems. Vol. 2. Scrivener Publishing LLC; 2021. pp. 193-217

[56] Essa FA, Abdullah AS, Omara ZM. Rotating discs solar still: New mechanism of desalination. Journal of Cleaner Production. 2020;275:123200

[57] Essa FA, Abou-Taleb FS, Diab MR. Experimental investigation of vertical solar still with rotating discs. Energy Sources, Part A: Recovery, Utilization, and Environmental Effects. 2021:1-21

[58] Kabeel A, et al. Experimental study on tubular solar still using Graphene Oxide Nano particles in Phase Change Material (NPCM's) for fresh water production. Journal of Energy Storage. 2020;28:101204

[59] Elashmawy M. Improving the performance of a parabolic concentrator solar tracking-tubular solar still (PCST-TSS) using gravel as a sensible heat storage material. Desalination. 2020;473:114182

[60] Essa FA, Abdullah AS, Omara ZM. Improving the performance of tubular solar still using rotating drum – Experimental and theoretical investigation. Process Safety and Environmental Protection. 2021;148:579-589

[61] Abdullah A, et al. Rotating-drum solar still with enhanced evaporation and condensation techniques: Comprehensive study. Energy Conversion and Management. 2019;199:112024

[62] Abdullah AS, et al. Experimental investigation of a new design of drum solar still with reflectors under different conditions. Case Studies in Thermal Engineering. 2021;24:100850

[63] Pounraj P, et al. Experimental investigation on Peltier based hybrid PV/T active solar still for enhancing the overall performance. Energy Conversion and Management. 2018;168:371-381

[64] Hedayati-Mehdiabadi E, Sarhaddi F, Sobhnamayan F. Exergy performance evaluation of a basin-type double-slope solar still equipped with phase-change material and PV/T collector. Renewable Energy. 2020;145:2409-2425

[65] Omara Z, Hamed MH, Kabeel A. Performance of finned and corrugated absorbers solar stills under Egyptian conditions. Desalination. 2011;277(1-3):281-287

[66] Abdullah A, et al. New design of trays solar still with enhanced evaporation methods–Comprehensive study. Solar Energy. 2020;203:164-174

[67] Abdullah AS, et al. Improving the trays solar still performance using reflectors and phase change material with nanoparticles. Journal of Energy Storage. 2020;31:101744

[68] Essa FA, et al. Experimental study on the performance of trays solar still with cracks and reflectors. Applied Thermal Engineering. 2021;188:116652

[69] Abdullah AS, et al. Improving the performance of trays solar still using wick corrugated absorber, nano-enhanced phase change material and photovoltaics-powered heaters. Journal of Energy Storage. 2021;40:102782

[70] Kumar PN, et al. Experimental investigation on the effect of water mass in triangular pyramid solar still integrated to inclined solar still. Groundwater for Sustainable Development. 2017;5:229-234

[71] Omara Z, Kabeel A, Essa F. Effect of using nanofluids and providing vacuum on the yield of corrugated wick solar still. Energy Conversion and Management. 2015;103:965-972

[72] Abdullah A, et al. Rotating-wick solar still with mended evaporation technics: Experimental approach. Alexandria Engineering Journal. 2019;**58**(4):1449-1459

[73] Omara ZM, et al. Experimental investigation of corrugated absorber solar still with wick and reflectors. Desalination. 2016;**381**:111-116

[74] Omara ZM, Kabeel AE, Essa FA. Effect of using nanofluids and providing vacuum on the yield of corrugated wick solar still. Energy Conversion and Management. 2015;**103**:965-972

[75] Modi KV, Nayi KH, Sharma SS. Influence of water mass on the performance of spherical basin solar still integrated with parabolic reflector. Groundwater for Sustainable Development. 2020;**10**:100299

[76] Abderachid T, Abdenacer K. Effect of orientation on the performance of a symmetric solar still with a double effect solar still (comparison study). Desalination. 2013;**329**:68-77

[77] Rajaseenivasan T, Kalidasa Murugavel K. Theoretical and experimental investigation on double basin double slope solar still. Desalination. 2013;**319**:25-32

[78] Rajaseenivasan T, Elango T, Kalidasa Murugavel K. Comparative study of double basin and single basin solar stills. Desalination. 2013;**309**:27-31

[79] El-Sebaii A. Thermal performance of a triple-basin solar still. Desalination. 2005;**174**(1):23-37

[80] Suneja S, Tiwari GN. Effect of water depth on the performance of an inverted absorber double basin solar still. Energy Conversion and Management. 1999;**40**(17):1885-1897

[81] Attia MEH, et al. Enhancement of hemispherical solar still productivity using iron, zinc and copper trays. Solar Energy. 2021;**216**:295-302

[82] Essa FA, et al. Experimental investigation of convex tubular solar still performance using wick and nanocomposites. Case Studies in Thermal Engineering. 2021;**27**:101368

[83] Omara ZM, et al. Experimental study on the performance of pyramid solar still with novel convex and dish absorbers and wick materials. Journal of Cleaner Production. 2021

[84] Alawee WH, et al. Improving the performance of pyramid solar still using rotating four cylinders and three electric heaters. Process Safety and Environmental Protection. 2021;**148**:950-958

[85] Essa F, et al. Enhancement of pyramid solar distiller performance using reflectors, cooling cycle, and dangled cords of wicks. Desalination. 2021;**506**:115019

[86] Al-Madhhachi H, Smaisim GF. Experimental and numerical investigations with environmental impacts of affordable square pyramid solar still. Solar Energy. 2021; **216**:303-314

[87] Kabeel AE, Omara Z, Essa F. Numerical investigation of modified solar still using nanofluids and external condenser. Journal of the Taiwan Institute of Chemical Engineers. 2017;**75**:77-86

[88] Khechekhouche A, et al. Energy, Exergy Analysis, and Optimizations of Collector Cover Thickness of a Solar Still in El Oued Climate, Algeria. International Journal of Photoenergy. 2021;**2021**:6668325

[89] Thamizharasu P, Shanmugan S, Gorjian S, et al. Improvement of Thermal Performance of a Solar Box Type Cooker Using SiO_2/TiO_2

Nanolayer. Silicon; 2020. https://doi.org/10.1007/s12633-020-00835-1

[90] Shanmugan S, et al. Experimental study on single slope single basin solar still using TiO_2 nano layer for natural clean water invention. Journal of Energy Storage. 2020;**30**:101522

[91] Arani RP, Sathyamurthy R, Chamkha A, et al. Effect of fins and silicon dioxide nanoparticle black paint on the absorber plate for augmenting yield from tubular solar still. Environ Sci Pollut Res. 2021;**28**;35102-35112. https://doi.org/10.1007/s11356-021-13126-y

[92] Yadav YP. Performance analysis of a solar still coupled to a heat exchanger. Desalination. 1991;**82**(1):243

[93] Valsaraj P. An experimental study on solar distillation in a single slope basin still by surface heating the water mass. Renewable Energy. 2002;**25**(4):607-612

[94] Modi KV, Shukla DL. Regeneration of liquid desiccant for solar air-conditioning and desalination using hybrid solar still. Energy Conversion and Management. 2018;**171**:1598-1616

[95] Panchal H, et al. Enhancement of the yield of solar still with the use of solar pond: A review. Heat Transfer. 2021;**50**(2):1392-1409

[96] Abdallah S, Abu-Khader MM, Badran O. Effect of various absorbing materials on the thermal performance of solar stills. Desalination. 2009;**242**(1-3):128-137

[97] Abdullah AS, et al. Enhancing the solar still performance using reflectors and sliding-wick belt. Solar Energy. 2021;**214**:268-279

[98] Omara Z, et al. Performance evaluation of a vertical rotating wick solar still. Process Safety and Environmental Protection. 2021; **148**:796-804

[99] Diab MR, Essa FA, Abou-Taleb FS, Omara ZM. Solar still with rotating parts: a review. Environ Sci Pollut Res Int. 2021 Oct;**28**(39):54260-54281. DOI: 10.1007/s11356-021-15899-8. Epub 2021 Aug 14. PMID: 34390475.

[100] Panchal H, et al. Experimental investigation on the yield of solar still using manganese oxide nanoparticles coated absorber. Case Studies in Thermal Engineering. 2021;**25**:100905

[101] Kabeel A, et al. Augmentation of a solar still distillate yield via absorber plate coated with black nanoparticles. Alexandria Engineering Journal. 2017;**56**(4):433-438

[102] Vigneswaran VS, et al. Augmenting the productivity of solar still using multiple PCMs as heat energy storage. Journal of Energy Storage. 2019;**26**:101019

[103] Abdelgaied M, et al. Improving the tubular solar still performance using square and circular hollow fins with phase change materials. Journal of Energy Storage. 2021;**38**:102564

[104] Thalib MM, et al. Comparative study of tubular solar stills with phase change material and nano-enhanced phase change material. Energies. 2020;**13**(15):3989

[105] Omara ZM, Hamed MH, Kabeel A. Performance of finned and corrugated absorbers solar stills under Egyptian conditions. Desalination. 2011;**277**(1-3):281-287

[106] Younes M, et al. Enhancing the wick solar still performance using half barrel and corrugated absorbers. Process Safety and Environmental Protection. 2021;**150**:440-452

[107] Hassan H. Comparing the performance of passive and active

double and single slope solar stills incorporated with parabolic trough collector via energy, exergy and productivity. Renewable Energy. 2020;**148**:437-450

[108] Abdallah S, Badran O. Sun tracking system for productivity enhancement of solar still. Desalination. 2008;**220**(1-3):669-676

[109] Al-Hinai H, Al-Nassri M, Jubran B. Effect of climatic, design and operational parameters on the yield of a simple solar still. Energy Conversion and Management. 2002;**43**(13): 1639-1650

[110] Abdullah A, et al. Experimental investigation of single pass solar air heater with reflectors and turbulators. Alexandria Engineering Journal. 2020; **59**(2):579-587

[111] Scrivani A, Bardi U. A study of the use of solar concentrating plants for the atmospheric water vapour extraction from ambient air in the Middle East and Northern Africa region. Desalination. 2008;**220**(1-3):592-599

[112] Essa F, et al. Extracting water content from the ambient air in a double-slope half-cylindrical basin solar still using silica gel under Egyptian conditions. Sustainable Energy Technologies and Assessments. 2020;**39**:100712

[113] Elashmawy M, Alshammari F. Atmospheric water harvesting from low humid regions using tubular solar still powered by a parabolic concentrator system. Journal of Cleaner Production. 2020;**256**:120329

[114] Gnanaraj SJP, Ramachandran S, Christopher DS. Enhancing the design to optimize the performance of double basin solar still. Desalination. 2017;**411**:112-123

[115] Badran OO. Experimental study of the enhancement parameters on a single

slope solar still productivity. Desalination. 2007;**209**(1-3):136-143

[116] Abad HKS, et al. A novel integrated solar desalination system with a pulsating heat pipe. Desalination. 2013;**311**:206-210

[117] El-Bahi A, Inan D. A solar still with minimum inclination, coupled to an outside condenser. Desalination. 1999;**123**(1):79-83

[118] Tanaka H. Experimental study of a basin type solar still with internal and external reflectors in winter. Desalination. 2009;**249**(1):130-134

[119] Tanaka H, Nosoko T, Nagata T. Experimental study of basin-type, multiple-effect, diffusion-coupled solar still. Desalination. 2002;**150**(2):131-144

[120] Yadav Y. Transient analysis of double-basin solar still integrated with collector. Desalination. 1989;**71**(2): 151-164

[121] Dayem AMA. Experimental and numerical performance of a multi-effect condensation–evaporation solar water distillation system. Energy. 2006; **31**(14):2710-2727

[122] Srithar K, Mani A. Open fibre reinforced plastic (FRP) flat plate collector (FPC) and spray network systems for augmenting the evaporation rate of tannery effluent (soak liquor). Solar Energy. 2007;**81**(12):1492-1500

[123] Shehabeldeen TA, et al. A novel method for predicting tensile strength of friction stir welded AA6061 aluminium alloy joints based on hybrid random vector functional link and henry gas solubility optimization. IEEE Access. 2020;**8**:79896-79907

[124] Elaziz MA, Elsheikh AH, Sharshir SW. Improved prediction of oscillatory heat transfer coefficient for a thermoacoustic heat exchanger using

modified adaptive neuro-fuzzy inference system. International Journal of Refrigeration. 2019;**102**:47-54

[125] Elsheikh AH, et al. An artificial neural network based approach for prediction the thermal conductivity of nanofluids. SN Applied Sciences. 2020;**2**(2):235

[126] Shehabeldeen TA, et al. Modeling of friction stir welding process using adaptive neuro-fuzzy inference system integrated with harris hawks optimizer. Journal of Materials Research and Technology. 2019;**8**(6):5882-5892

[127] Babikir HA, et al. Noise prediction of axial piston pump based on different valve materials using a modified artificial neural network model. Alexandria Engineering Journal. 2019; **58**(3):1077-1087

[128] Elsheikh AH, et al. Modeling of solar energy systems using artificial neural network: A comprehensive review. Solar Energy. 2019;**180**:622-639

[129] El-Hawary M. Artificial neural networks and possible applications to desalination. Desalination. 1993; **92**(1-3):125-147

[130] Zarei T, Behyad R. Predicting the water production of a solar seawater greenhouse desalination unit using multi-layer perceptron model. Solar Energy. 2019;**177**:595-603

[131] Tayyebi S, Alishiri M. The control of MSF desalination plants based on inverse model control by neural network. Desalination. 2014;**333**(1): 92-100

[132] Cao W, et al. Modeling and simulation of VMD desalination process by ANN. Computers and Chemical Engineering. 2016;**84**:96-103

[133] Aish AM, Zaqoot HA, Abdeljawad SM. Artificial neural network approach for predicting reverse osmosis desalination plants performance in the Gaza Strip. Desalination. 2015; **367**:240-247

[134] Santos NI, et al. Modeling solar still production using local weather data and artificial neural networks. Renewable Energy. 2012;**40**(1):71-79

[135] Alhamdoosh M, Wang D. Fast decorrelated neural network ensembles with random weights. Information Sciences. 2014;**264**:104-117

Generating Artificial Weather Data Sequences for Solar Distillation Numerical Simulations

Bao The Nguyen

Abstract

According to the natural geographical distribution, developing countries are concentrated in tropical climates, where radiation is abundant. So the use of solar energy is a sustainable solution for developing countries. However, daily or hourly measured solar irradiance data for designing or running simulations for solar systems in these countries is not always available. Therefore, this chapter presents a model to calculate the daily and hourly radiation data from the monthly average daily radiation. First, the chapter describes the application of Aguiar's model to the calculation of daily radiation from average daily radiation data. Next, the chapter presents an improved Graham model to generate hourly radiation data series from monthly radiation. The above two models were used to generate daily and hourly radiation data series for Ho Chi Minh City and Da Nang, two cities representing two different tropical climates. The generated data series are tested by comparing the statistical parameters with the measured data series. Statistical comparison results show that the generated data series have acceptable statistical accuracy. After that, the generated radiation data series continue to be used to run the simulation program to calculate the solar water distillation system and compare the simulation results with the radiation data. Measuring radiation. The comparison results once again confirm the accuracy of the solar irradiance data generation model in this study. Especially, the model to generate the sequences of hourly solar radiation values proposed in this study is much simpler in comparison to the original model of Graham. In addition, a model to generate hourly ambient tempearure date from monthly average daily ambient temperature is also presented and tested. Then, both generated hourly solar radiation and ambient temperature sequences are used to run a solar dsitillation simulation program to give the outputs as monthly average daily distillate productivities. Finally, the outputs of the simulation program running with the generated solar radiation and ambient temperature data are compared with those running with measured data. The errors of predicted monthly average daily distillate productivities between measured and generated weather data for all cases are acceptably low. Therefore, it can be concluded that the model to generate artificial weather data sequences in this study can be used to run any solar distillation simulation programs with acceptable accuracy.

Keywords: solar stills, solar distillation numerical simulations, monthly average daily solar radiation, hourly solar radiation values, monthly average daily temperature, hourly monthly-average ambient temperature, Markov Transition Matrix

1. Introduction

With the development of computers, simulation programs are increasingly developed and become useful and indispensable tools for researchers and designers. It helps users to optimize design and system parameters without having to spend money to build experimental models and waste time to conduct experiments. Pro-phylactic programs in the field of solar water distillation are no exception. To run solar water distillation simulations, users need to provide weather data such as solar irradiance and ambient temperature measured in days, hours or smaller time periods. According to the requirements of the software. However, hourly weather data is not always available, especially in developing countries because measuring hourly weather data requires equipment, time and money. According to Duffie and Beckman [1], these weather data must be collected for at least 8 years to get the average value to remove the anomalies of the weather such as El Nino, La Nina phenomena, etc.

Another solution is to use typical meteorological year (TMY) data. In fact, the concept of TMY is derived from long-term weather data, which is determined in the correlation and statistical distribution to determine the characteristic indexes to produce the average value [2]. These data are then extracted from the selection criteria to produce month-by-month data from 23 years' data. TMY data were established for 26 Canadian sites, and were applied to the concept of a similar test reference year (TRY) for Europe [2]. While this approach reduces computational effort and the data base required to run simulations, the metric is also based on long-term data, something not available in most places in the world, especially in developing countries.

As pointed out by Nguyen and Hoang [3], the shortage of weather data, especially solar radiation data is very serious in developing countries. For example, in Vietnam, out of a total of 171 hydro-meteorological stations, only 12 have total solar radiation data, of which only 9 have continuous measurements. The remaining meteorological stations only record the number of hours of sunshine. Furthermore, radiation metrics are manually measured by humans every 3 hours instead of hourly. Therefore, hourly radiation data at hydro-meteorological stations in this country are not reliable enough to be used for simulation programs using solar energy systems.

There are two ways to solve the problem of lack of measurement data at the survey site: (i) using extrapolation to process data from hydrometeorological sites adjacent to similar climate features, and (ii) use aggregate generation to generate a series of weather data from the data requiring at least monthly averages. However, the first method can lead to large errors, moreover, very few developing countries have such data available [2]. Therefore, the following method has been studied and developed by many researchers.

Many researchers have proposed mathematical models to calculate the complete series of weather data. Fernandez-Peruchena et al. [4] and Boland [2] used numer-ical methods to probabilistically simulate daily and hourly solar irradiance data series. Brecl and Topic [5] used a similar approach to generate daily and hourly solar irradiance data from average daily irradiance values. Bright et al. [6] and Hofmann et al. [7] also apply statistical probabilistic techniques to generate a series of solar irradiance values per minute or every 5 minutes from hourly solar irradiance data. Soubdhan and Emilion [8] even used a random method to generate a sequence of solar radiation in seconds. Magnano et al. [9] applied the same technique to gener-ate a synthetic sequence of half an hour's temperature. A common feature of the aforementioned studies is the use of a probability distribution function (PDF) of the data to normalize random variables to bring them to a Gaussian distribution [10].

Gafurov et al. [11] another approach was to incorporate the spatial correlation of solar irradiance (SCSR) into random solar irradiance data generation models to generate monthly solar irradiance time series and daily.

Recently, several researchers have used different types of artificial neural networks (ANNs) to model the values of total solar irradiance on horizontal surfaces, such as [12–16]. Wu and Chan [17] used a new combined model of ARMA (Automatic Recovery and Moving Average) and TDNN (Time Delayed Neural Network) to predict hourly solar irradiance in Singapore. However, Mora-Lopez [10] pointed out that the limitation of these methods is that they are "black boxes" for outputting and analyzing averages of daily global irradiance, resulting in no important information can be obtained from these methods. Mora-Lopez et al. [18] proposed to use machine learning theory with a combination of probability finite automata (PFA) to calculate the values of total daily solar irradiance. The limitation of this method is that the use of PFA is complex and the method has not been shown to be universally applicable.

The results of the above review and analysis show that the stochastic methods are still globally applicable, simple and require minimal input data. Therefore, in this study, randomization technique was chosen to generate series of weather data, including solar irradiance and daily and hourly ambient temperature from monthly averages. These are important weather metrics for running the numerical simulations of solar distillation systems. First, a stochastic model is used to generate a composite of daily irradiance from monthly average daily solar irradiance values. The generated daily radiation sequences are then used to generate the hourly solar radiation sequences. Similarly, a model for generating hourly temperature series from monthly mean temperature values is also presented.

2. Stochastic model for daily radiation data generation

2.1 Model of generating daily solar radiation sequence

When analyzing data of 300 months of solar radiation taken from 9 hydro-meteorological stations with different weather characteristics, Aguiar et al. [19] discovered the analyzed solar irradiance values in For any time period, there is a probability distribution function that seems to be related to the monthly mean clearness index, $\overline{K}_T$, for that time period. Furthermore, they also found that the daily radiation value of any given day is statistically related to the value of the previous day. Based on this finding, Aguiar and colleagues built up 10 Markov matrices (called MTM library) from the data analysis of 300 months of solar radiation mentioned above. 10 subdivided matrices include: 1 matrix with $\overline{K}_T \leq 0.3$ typical for months with very low direct radiation components; The next 8 matrices are for the months where $\overline{K}_T$ varies from 0.3 to 0.7 with the increment of $\overline{K}_T$ increasing by 0.05 for the next matrix; the final matrix with $\overline{K}_T > 0.7$ is for months with a high direct radiation component. The MTM library was then used to generate the daily radiation series from the average daily irradiance values for the locations in the United States for which the irradiance data were not used to generate the aforementioned MTM library. This simulation result was compared with the measured radiation results and compared with the simulation results from Graham's model [20]. When comparing statistical parameters such as mean, variance, and probability density functions as well as statistic characteristics (e.g. autocorrelation functions), Aguiar's model produced more accurate data series than with the Graham model. Furthermore, Aguiar's model is computationally simpler than Graham's model [21].

To calculate the K_T values from the monthly mean $\overline{K}_T$, in this study, the Aguiar method was chosen for the following reasons:

- The calculation expressions given by Graham are based on results that are built from the data of the United States, so certainly not suitable for tropical climates [22]. This was confirmed by the Nguyen and Pryor [21] in their study, and even in high K_T regions, the above expressions do not fit the curve due to Liu and Jordan [21].

- The above disadvantage of Graham's method can be overcome by adding the expressions for tropical climates developed by Aguiar group [19], but the use of this expression has not been fully verified by scientists and will create a complex, climate-dependent computational model.

- The locations where the Aguiar group used the measured data to build the MTM matrices include many characteristic climate zones, in these locations there is one location with a central tropical rainforest (C_{aw}) climate, which is Macao. and one location with a rainforest climate (A_w) is Polana, Mozambique. The use of this method would therefore be suitable for the study's objective of tropical climates, where most developing countries are often located.

2.2 The data are used to evaluate the accuracy of the model

Solar irradiance data were measured in 2 cities representing 2 tropical climates to evaluate the accuracy of daily solar radiation data generated from the calculated model selected in this study. They are Ho Chi Minh City representing the tropical forest climate (A_w) and Da Nang representing the tropical monsoon climate (A_m). Pyranometers were used to measure total irradiance on a horizontal plane in the two cities every 5 minutes, measuring continuously from 5:30 a.m. to 6:30 p.m. Since these two cities have low latitudes (10^0N and 16^0N respectively), the day length does not change much during the year, so there is no need to extend the seasonal solar irradiance measurement period of the year [3]. Then, solar irradiance by hour, by day and by day of month average is calculated from the measured data. **Table 1** presents the average daily solar irradiance of the two cities mentioned above, used to run the program to generate date and time irradiance data in this study.

From the series of daily radiation values of 365 days of the year, the series of 365 values of the daily clearness index is calculated according to the following Eq. (1):

$$K_{Tmea.} = \frac{H}{H_0} \tag{1}$$

where H is the total daily solar radiation measured on a horizontal plane and H_0 is the daily radiation outside the atmosphere, calculated by the equation:

	Jan.	Feb.	Mar.	Apr.	May	Jun.	Jul.	Aug.	Sep.	Oct.	Nov.	Dec.
HCM	13.0	18.0	18.1	18.7	16,7	17,4	17.3	17.6	15.9	15.0	14.7	13.9
Danang	10.3	18.8	18.6	22.1	22.9	23.9	20.3	18.7	17.2	14.8	11.9	8.4

Table 1.
Total monthly average solar irradiance measured on a horizontal plane in Ho Chi Minh City and Da Nang (MJ/m². day).

$$H_0 = \frac{24}{\pi} G_{SC} 3600 \left\{ \left[1 + 0,033. \cos\left(\frac{360n}{365}\right) \right] \right.$$
$$\left. \times \left[\cos\phi \cos\delta \sin\omega_S + \frac{\pi}{180}\omega_S \sin\phi \sin\delta \right] \right\} \qquad (2)$$

with G_{sc}, n, ϕ, δ and ω_s respectively are solar constance, day of the year, latitude of the investigated location, declination angle and sunset hour angle, defined in [1].

From the values of daily clearness index K_T, the monthly average daily values of clearness index $\overline{K}_T$ for 12 months of the year are calculated:

$$\overline{K_T} = \frac{\overline{H}}{\overline{H}_o} \qquad (3)$$

where the monthly average daily irradiance values $\overline{H}$ are taken in **Table 1** and monthly average daily irradiance values outside the atmosphere $\overline{H}_0$ are calculated by Eq. (2) with day n being the average day of the month, given in [1]. **Table 2** presents the $\overline{K}_T$ values of the 2 investigated cities.

2.3 Applying Aguiar's model

Figure 1 presents the procedure for calculating the series of daily clearness index from the monthly average daily values.

After 365 values of the daily photometric index are calculated for each location, these value series are compared with the measurement series through statistical functions such as cumulative distribution function (CDF), density function, etc. probability (PDF). **Figures 2** and **3** present the cumulative distribution function of CDF of the calculated and measured K_T in Ho Chi Minh City and Da Nang while **Figures 4** and **5** represent the probability density function PDF for these two cities. Statistical parameters including mean, median, minimum, maximum, standard deviation, mean absolute error (MAE) and mean square error (RMSE) were also compared between the K_T series. Calculated and measured, as shown in **Tables 3** and **4**.

The results shown in **Figures 2–5** show that the Aguiar model produced a K_T daily value series with an acceptable level of accuracy compared with the measured series. Similarly, the statistical parameters in **Tables 3** and **4** also show that the statistical error between the calculated and measured series is relatively small. Specifically, the mean and median error percentages of generated chains are 1% and − 4% for Ho Chi Minh City and 6% and 14% for Da Nang, respectively. Therefore, this model is expected to be able to be used to generate a series of daily cloud optical coefficients for any location because the Aguiar model has been proven to be universally applicable in the world [2, 4, 5, 19, 22]. As shown above, this model only requires input of 12 average daily solar irradiance values at the location to be calculated.

	Jan.	Feb.	Mar.	Apr.	May	Jun.	Jul.	Aug.	Sep.	Oct.	Nov.	Dec.
HCM	0.42	0.53	0.50	0.50	0.45	0.47	0.47	0.47	0.44	0.42	0.47	0.46
Danang	0.36	0.59	0.53	0.59	0.60	0.63	0.53	0.50	0.48	0.45	0.40	0.30

Table 2.
Monthly average daily values of clearness index in Ho Chi Minh City and Da Nang.

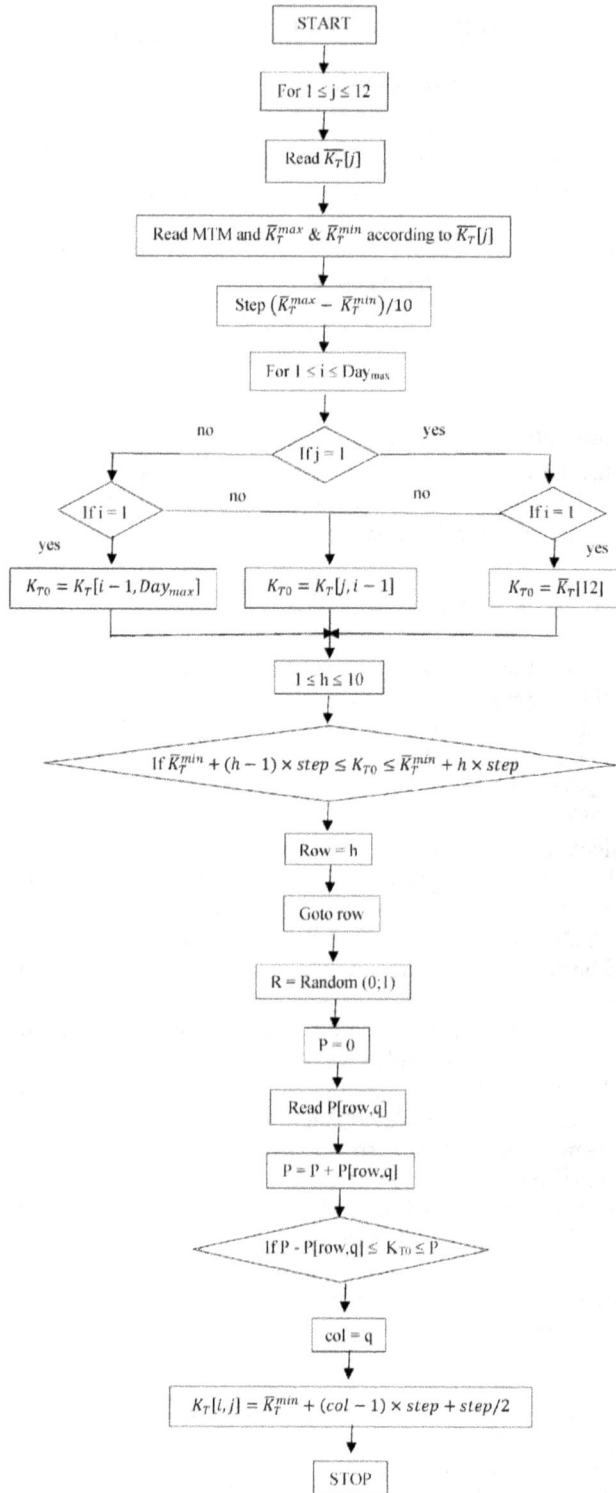

Figure 1.
The procedure for calculating the series of K_T from $\overline{K}_T$.

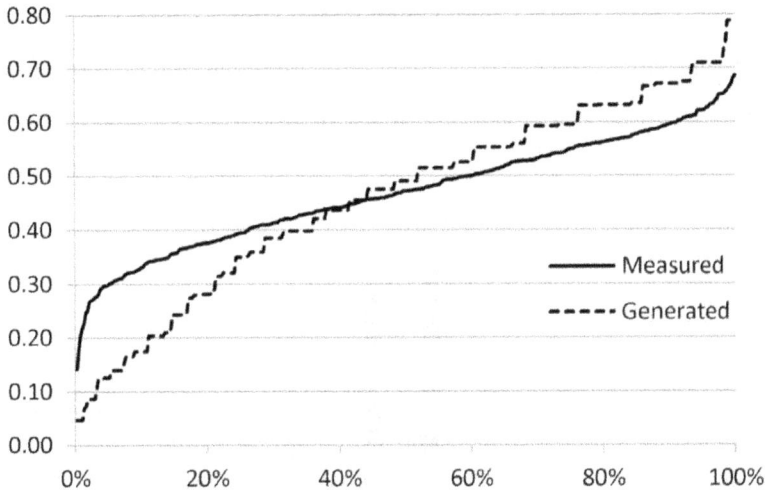

Figure 2.
Cumulative distribution function of K_T for Ho Chi Minh City.

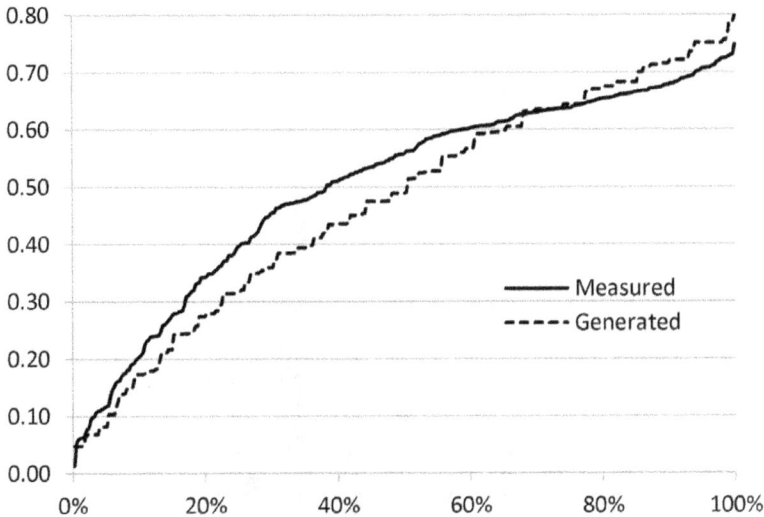

Figure 3.
Cumulative distribution function of K_T for Da Nang.

3. Model of generating hourly solar radiation

3.1 Choosing a model to create hourly solar radiation chains

There are many studies in the world on establishing mathematical models to generate daily and hourly radiation data series [2, 4, 5]. Basically, these studies are based on the approach of Aguiar [19] and Graham [20].

With the assumption that the sky clarity index k_t depends only on the cloudiness coefficient of the K_T day, Graham et al. [20] analyzed k_t into two components: an average (or trend) component and a random component.

$$k_t = k_{tm} + \alpha \qquad (4)$$

The formula for calculating the trend component or the regular component:

$$k_{tm} = \lambda + \rho \, \exp.(-\kappa m) \qquad (5)$$

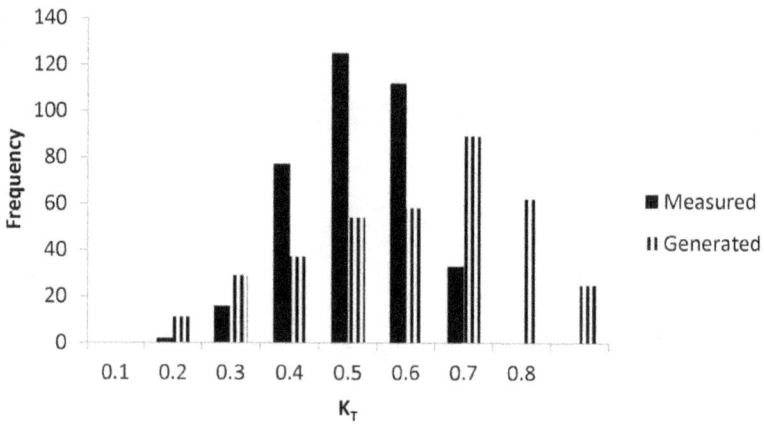

Figure 4.
Probability density function of K_T for Ho Chi Minh City.

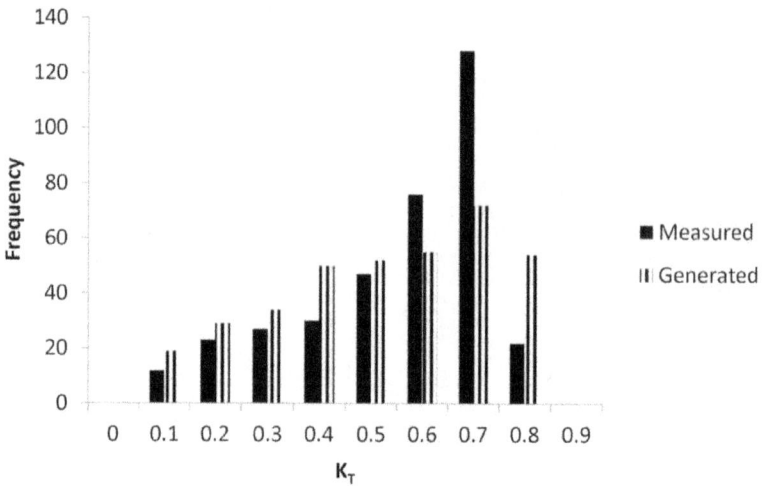

Figure 5.
Probability density function of K_T for Da Nang.

	Mean	Median	Min	Max	Std. dev.	MAE (%)	RMSE (%)
K_{Tmea}	0.47	0.47	0.14	0.69	0.10		
$K_{Tgen.}$	0.46	0.49	0.05	0.78	0.18		
Error (%)	1	−4				16.8	8.1

Table 3.
Statistic parameters of K_T series of Ho Chi Minh City.

	Mean	Median	Min	Max	Std. dev.	MAE (%)	RMSE (%)
K_{Tmea}	0.50	0.56	0.01	0.75	0.18		
$K_{Tgen.}$	0.47	0.49	0.05	0.80	0.20		
Error (%)	6	14				4.5	5.1

Table 4.
Statistic parameters of K_T series of Da Nang.

where m is the air mass, the value calculated at the time of the middle of the hour. The parameters λ, ρ and κ are the identity function of K_T:

$$\lambda(K_T) = K_T - 1.167 K^3{}_T(1-K_T) \tag{6}$$

$$\rho(K_T) = 0.979(1-K_T) \tag{7}$$

$$\kappa(K_T) = 1.141(1-K_T)/K_T \tag{8}$$

The standard deviation σ_α of the variable α is expressed as:

$$\sigma_\alpha(K_t) = 0.16 \sin(\pi K_t/0.90) \tag{9}$$

Then use the Gaussian normalization technique to transform this variable α into a Gaussian variable β with the relation between α and β as follows:

$$\alpha = -\frac{1}{1.158} \ln \left\{ \frac{1}{0.5\left[1 + erf\left(\frac{\beta}{\sqrt{2}}\right)\right]} - 1 \right\} \tag{10}$$

Then apply ARMA models to the data series β and determine that β follows the model as AR(1):

$$\beta_t = \phi\beta_{t-1} + \vartheta_t \tag{11}$$

where:
β_{t-1} is the value of the variable at t-1.
Φ is the automatic regression coefficient.
ϑ_t is a random number from a normal distribution with zero mean and a standard deviation $\sqrt{1-\phi}$
As in the case of using date values, the coefficient Φ varies slightly by locality but the value 0.54 can be chosen as the value to use in the model for all localities.
Aguiar's method [23] to generate k_T is similar to Graham's method but has some differences as follows:

- First, the standard deviation σα depends not only on K_T but also on the altitude angle of the sun h_s:

$$\sigma_\alpha(K_T, h_s) = A \times exp\{B \times [1 - \sin(h_s)]\} \tag{12}$$

with:

$$A = 0.14 \times exp\left[-20.0(K_T - 0.32)^2\right] \tag{13}$$

$$B = 3.0(K_T - 0.45)^2 + 16.0K_T^5 \tag{14}$$

- Second, the coefficient Φ depends on K_T according to the following expression:

$$\Phi = 0.38 + 0.06 \cos (7.4K_T–2.5) \tag{15}$$

- Third, the calculated k_t values are limited by the clear sky clearness index k_{cs}.

$$k_{cs}(t) = 0.88 \times \cos [\pi(t - 12.5)/30] \tag{16}$$

With t is the hour considered.

Aguiar's model has been successfully applied to generate hourly solar irradiance in Spain and Slovenia [4, 5], while Graham's approach to generate hourly solar irradiance sequences at different locations has been successfully applied for locations in the United States in particular and many parts of the world in general [24]. On the other hand, Aguiar and Graham's models were used in generating and comparing hourly solar irradiance for 6 locations in Australia and the results showed that Graham's model produces a global solar irradiance sequence which better fit for all six sites [21]. Therefore, Graham's model was chosen to generate hourly solar radiation in this study.

However, Graham's model also has some disadvantages as follows:

- The use of Gaussian mapping technique to process the random component of kt values is very complicated and time consuming.

- The error of the hourly series of transparency index generated from Graham's model in comparision to the measured series is larger than the error from the model sugeeted in this study.

To solve the complicated and time consuming problem of the Graham model, the β values given in Eq. (10) are transformed to a non-standard distribution h by using Norminv function in MATLAB. The random components of the k_t value are then calculated by:

$$\alpha = h \times \sigma_\alpha (K_T) \tag{17}$$

with $\sigma_\alpha (K_T)$ is the standard deviation computed by using Eq. (9).

This suggested model are not only much simpler than Graham's method, but also produces more accurate results. **Table 5** demonstrates the hourly solar irradiance sequence error generated by Graham's model and this study's modified model compared with the measured solar irradiance sequence for Ho Chi Minh City over 20 times run a program written in MATLAB.

3.2 Modified Graham model to create a series of hourly clearness indices

Figure 6 shows the process of creating a series of hourly clearness index series. This procedure is modified from the Graham model, as analyzed above. In this Figure:

ϕ is the investigated location's latitude

L_{st} and L_{loc} are respectively the longitude of the standard meridians and the considered location.

j is the month of the year.

i is the day of a month

ω_s is the angle of sunset for the calculated day

K_T [i] [j] is the daily clearness index of the ith day in the jth month

Generating program runtime	Errors of the generated versus measured hourly solar radiation values (%)	
	Graham's model	Modified model of this study
1	2.87%	0.74%
2	5.31%	2.70%
3	5.75%	2.49%
4	3.81%	1.46%
5	11.15%	9.47%
6	5.04%	2.84%
7	4.22%	2.06%
8	5.19%	3.60%
9	4.12%	2.27%
10	4.70%	2.18%
11	6.40%	4.40%
12	7.33%	5.29%
13	6.32%	4.22%
14	5.07%	2.60%
15	5.03%	2.78%
16	7.52%	4.67%
17	4.88%	2.52%
18	3.24%	1.48%
19	8.15%	6.44%
20	8.68%	5.36%

Table 5.
Errors of the generated versus measured hourly solar radiation values in Ho Chi Minh City.

ω is the calculated hour angle.

k_{tm} is the "long-term" average value of k_t

σ_{kt} is the standard deviation of k_t toward the values of the "long-term" average value

ε_t is a Gaussian distribution's random number

hr. is the investigated hour.

χ is a Gaussian distribution's random variable with "0" mean and "1" variance.

θ_1 is the parameter of the AR1 model.

F_{normal} is a function to convert a Gaussian variable into a non-normally distributed variable

MATLAB is used to write generation program for hourly k_t series.

3.3 Validate generated hourly clearness index strings

The daily generated transparency index values were used as input to the hourly k_t series generation program. The calculated k_t values are then compared with k_{tmea} values, where k_{tmea} is the measured hourly clearness index values given by:

$$k_{tmea.} = \frac{I}{I_0} \qquad (18)$$

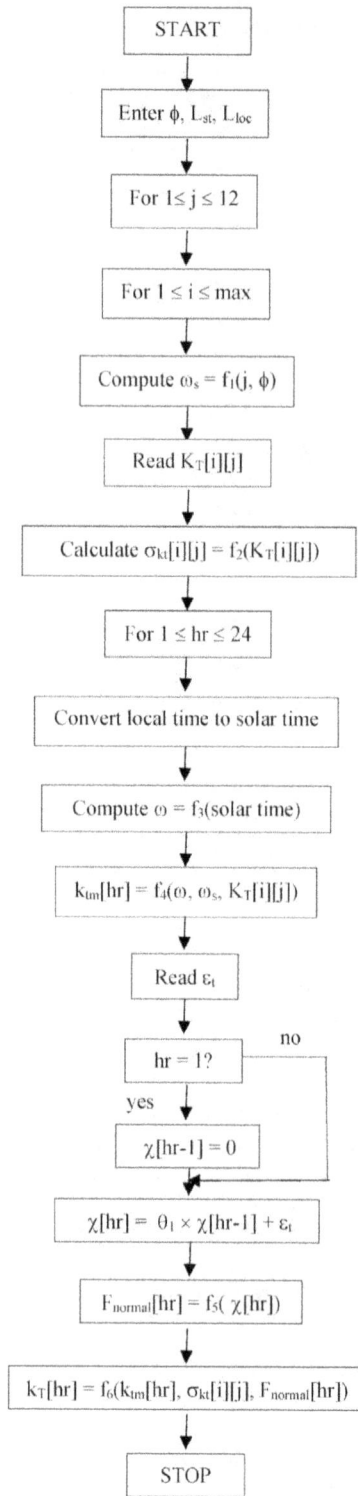

Figure 6.
Flow chart for generating hourly k_t series from daily optical clearness index series K_T.

where I is the horizontal total solar irradiance measured from ω_1 to ω_2 in Ho Chi Minh City and Da Nang; I_0 is the solar radiation outside the atmosphere from hour angle ω_1 to ω_2, given by [1]:

$$
\begin{aligned}
I_0 = \frac{12}{\pi} G_{SC} \times 3600 &\left\{ \left[1 + 0,033. \cos\left(\frac{360n}{365}\right) \right] \right. \\
&\times \left. \left[\cos\phi\cos\delta(\sin\omega_2 - \sin\omega_1) + \frac{\pi}{180}(\omega_2 - \omega_1)\sin\phi\sin\delta \right] \right\}
\end{aligned} \tag{19}
$$

The cumulative distribution function (CDF) graphs of k_t over the hours for Ho Chi Minh City and Da Nang are shown in **Figures 7** and **8** respectively while the probability density function (PDF) of k_t over the hour for these cities are shown in **Figures 9** and **10**. Additionally, several statistical parameters, including mean, median, minimum, maximum, standard deviation, mean absolute error (MAE) and mean square error (RMSE) of the measured and generated k_t series in Ho Chi Minh City and Da Nang are also shown in **Tables 6** and **7**, respectively.

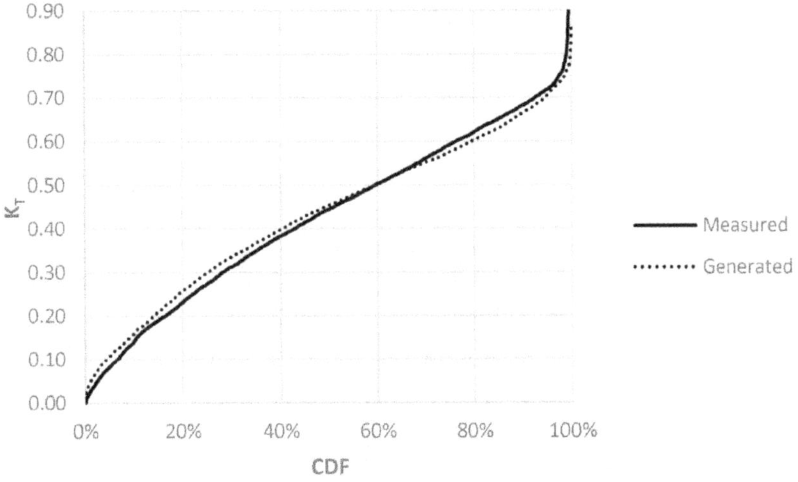

Figure 7.
Cumulative distribution function (CDF) of hourly k_t for Ho Chi Minh City.

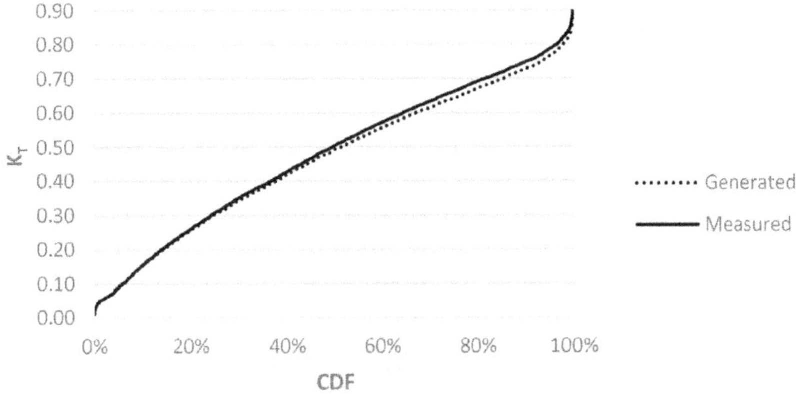

Figure 8.
Cumulative distribution function (CDF) of hourly k_t for Da Nang.

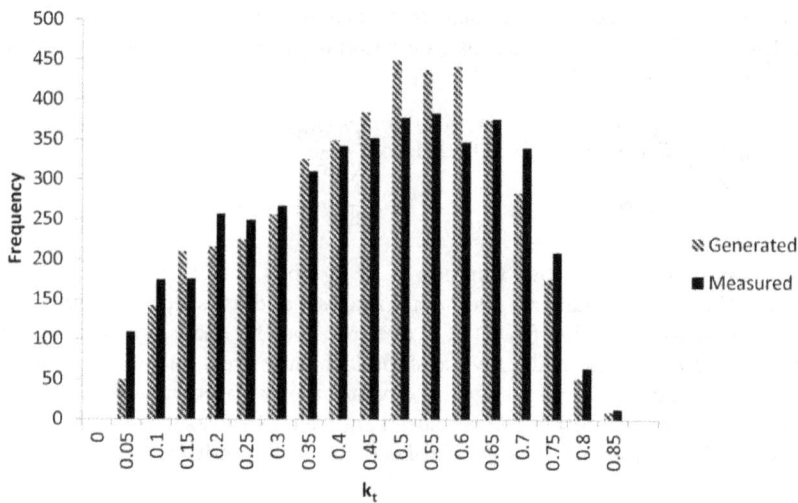

Figure 9.
Probability density function (PDF) of hourly k_t for Ho Chi Minh City.

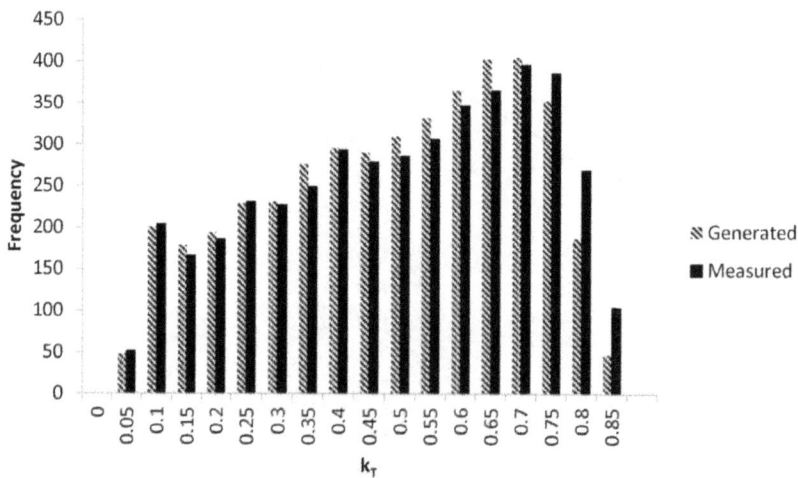

Figure 10.
Probability density function (PDF) of hourly k_t for Da Nang.

	Mean	Median	Min	Max	Std. dev.	MAE (%)	RMSE (%)
k_{tmea}	0.426	0.443	0.001	0.899	0.197		
$k_{tgen.}$	0.433	0.453	0.007	0.858	0.191		
Error (%)	−1.5	−2.4				2.0	0.03

Table 6.
Statistical parameters of the k_t series of Ho Chi Minh City.

As presented in **Figures 7–10** and **Tables 6** and **7**, the hourly k_t series for the two investigated cities have been successfully generated by the suggested model with very high accuracy. The mean and median error percentages of generated sequences were − 1.5% and − 2.4% for Ho Chi Minh City and − 1.3% and 0.3% for Da Nang,

	Mean	Median	Min	Max	Std. dev.	MAE (%)	RMSE (%)
k_{tmea}	0.459	0.491	0.019	0.905	0.210		
$k_{tgen.}$	0.465	0.492	0.012	0.884	0.217		
Error (%)	−1.3	−0.3				3.3	0.03

Table 7.
Statistical parameters of the k_t series of Da Nang.

respectively. Since stochastic models have been approved to have universal characteristics, as mentioned above, the model in this study is expected to be applicable to any location in the world.

4. Model to generate hourly ambient temperature sequences

The procedure to generate hourly ambient temperature sequences from monthly mean ambient temperature, $\overline{T}_a$, and monthly mean daily clearness index, $\overline{K}_t$, was described by Knight et al. [25]. The model to generate artificial hourly ambient temperature sequences for Australia was developed by Nguyen and Pryor [21].

First, to generate the deterministic component of the hourly ambient temperatures series, the concept of the average normalized diurnal temperature variation, developed by Erbs et al. [26], is applied. Hourly measured ambient temperature data from two locations (Ho Chi Minh city and Da Nang) are used to calculate the hourly monthly-average ambient temperature, $\overline{T}_{a,h}$, at each hour of the day for each month. These curves are standardized by subtracting the monthly-average daily temperature, $\overline{T}_a$, from each of the hourly values and then dividing by the amplitude of the curve (defined as the difference between the maximum and minimum hourly average temperatures over the day), A. Subsequently twelve cosine curves are derived for each location.

The average of these 24 (ie., 12 curves * 2 locations) standardized curves are calculated. Interestingly, the equation originally derived by Erbs and his colleagues is found to fit the average standardized curve in this study. The equation is expressed:

$$\frac{\overline{T}_{a,h} - \overline{T}_a}{A} = 0.4632 \cos{(t* - 3.805)} + 0.0984 \cos{(2\,t* - 0.360)}$$
$$+ 0.0168 \cos{(3\,t* - 0.822)} + 0.0138 \cos{(4\,t* - 3.513)} \quad (20)$$

$t*$ is given by: $t* = \frac{2\pi(t-1)}{24}$ where temperature is in hours with 1 and 24 corresponding to 1 am and midnight, respectively.

The relation between the amplitude A and the monthly mean clearness index, $\overline{K}_t$, is calculated as [26]:

$$A = 20.231\,\overline{K}_t - 3.103 \quad (21)$$

After the trend components are removed from the ambient temperature values, the variable component of the data is converted into a normal distribution and then tested with many ARMA models. The AR2 model is finally selected:

$$X_t = \Phi_1 X_{t-1} + \Phi_2 X_{t-2} + \varepsilon_t \quad (22)$$

here, χ_{t-1} and χ_{t-2} are the values of the weather data variables at t-1 and t-2, respectively, and Φ_1 and Φ_2 are calculated from the available ambient temperature data and are found to be 0.9072 and $-$ 0.1430, respectively. ε_t is a random number from a normal distribution with zero mean and a standard deviation $\sqrt{1 - \Phi_1 r_1 - \Phi_2 r_2}$. r_1 and r_2 are the corresponding autocorrelation coefficients.

The generated χ is transformed to an hourly temperature by equating the cumulative function of χ, F_{normal}, and hourly ambient temperature, F_{temp}. F_{normal} is given by:

$$F_{normal} = \frac{1}{\sqrt{2\pi}} \int_{-\infty}^{x} exp\left(-\frac{1}{2}t^2\right) dt = \frac{1}{2}1 + erf\left(\frac{\chi}{\sqrt{2}}\right) \tag{23}$$

whereas F_{temp} is calculated as follows:

$$F_{temp} = \frac{1}{1 + exp\left(-3.396h\right)} \tag{24}$$

with:

$$h = (T - \overline{T}_{a,h})\left(\sigma_m \sqrt{N_m/24}\right) \tag{25}$$

in which N_m is the number of hours in the months and σ_m is the standard deviation of the monthly-average daily temperature $\overline{T}_a$ given by:

$$\sigma_m = 1.45 + 0.029\overline{T}_a + 0.0664\sigma_{yr} \tag{26}$$

σ_{yr} is the standard deviation of the yearly average ambient temperature. Solving these equations, hourly ambient tempaerature is given by:

$$T = \overline{T}_a - \frac{\sigma_m \sqrt{\frac{N_m}{24}}}{3.369} \times \ln\left\{\frac{1}{0.5\left[1 + erf\left(\frac{\chi}{\sqrt{2}}\right)\right]} - 1\right\} \tag{27}$$

Figure 11 shows the schematic diagram of the procedure to generate hourly ambient temperature sequences.

In this figure:

L_{st} and L_{loc} are the standard meridian for the local time zone and the longitude of the location considered, respectively.

T_a [j] and K[j] are the monthly mean ambient temperature and monthly mean daily radiation of jth month, respectively.

$T_{a,yr}$ is the year average ambient temperature, calculated from the 12 monthly values.

σ_m [j] is the monthly standard deviation of the jth month, obtained from the yearly average value and the monthly average temperature for that month.

A[j] is the amplitude of the diurnal variation (peak to peak) of ambient temperature, and is a function of monthly average daily clearness index.

$T_{a,h}$ [avhr] is the hourly monthly-average ambient temperature; the subscript "avhr" indicates the calculated monthly-average hour (avhr = 1 to 24).

ε_t is a random number from a Gaussian distribution.

hr is the hour considered.

Nmax is the number of hours in the respective month

χ is the normally distributed stochastic variable with a mean of 0 & a variable of 1.

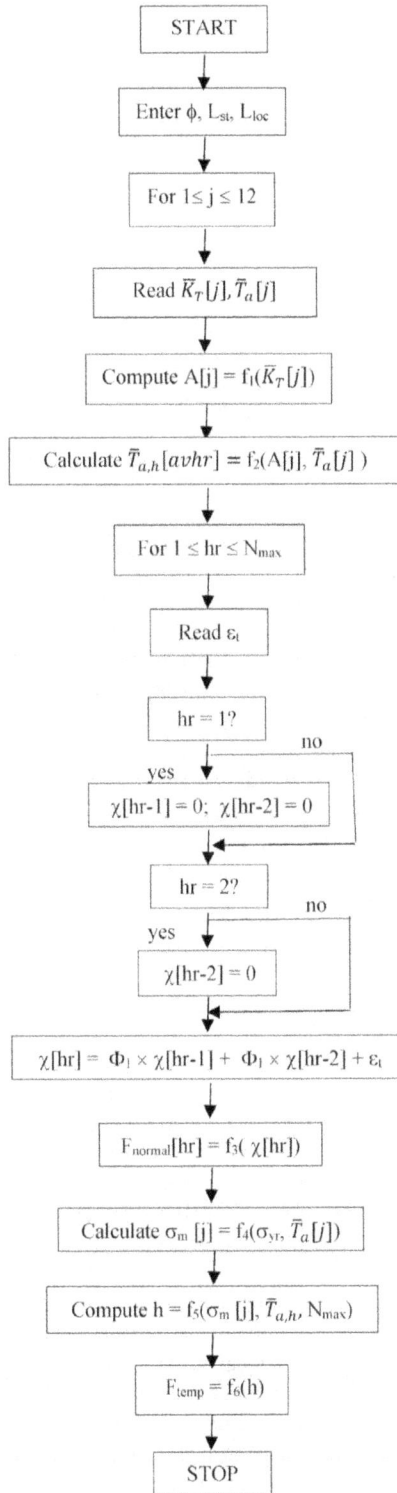

Figure 11.
Schematic diagram of the procedure to generate hourly ambient temperature sequences.

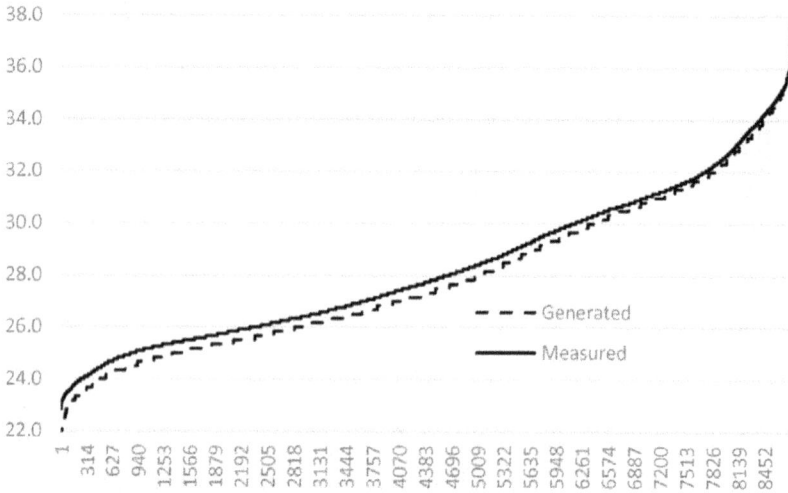

Figure 12.
Cumulative distribution function (CDF) of hourly ambient temperature for Ho Chi Minh City.

Φ_1 and Φ_2 are the first and second parameters of the AR2 model.

F_{normal} is the cumulative distribution of a normally distributed variable.

Figure 12 shows the cumulative distribution of hourly ambient temperature sequences for Ho Chi Minh City. The figure compares the results using measured data and arificial data generated from the equations described in this section. As shown, the model presented here produced accurate hourly ambient temperature in comparision with measured data.

5. Validating generated versus measured weather data by running a solar distillation simulation program

The main objective of this study is to build a model to generate weather data, including daily and hourly solar radiation sequences and ambient temperature series; then these weather data chains must be used to run simulation programs. Therefore, the generated weather data is used as input for SOLSTILL – a simulation program for solar distillation systems [27]. This simulation program was designed to enable to simulate both passive solar stills and active solar distillation systems. **Figure 13** presents the heat and mass diagrams in a passive solar still whereas **Figures 14** and **15** respectively shows the schematic diagram and heat and mass transfer process in a forced circulation solar still.

The inputs for SOLSTILL can be in form of hourly weather data if the measured hourly solar radiation and ambient temperature are available. If not, the weather data generation function in SOLSTILL can be called to generate weather sequences with input as monthly average daily solar radation and ambient temperature values of 12 months [27]. Both modes of weather data in SOLSTILL (i.e., input hourly weather data and generation mode) are used in this study. For Mode 1, hourly measured weather values, achieved from National Center for Hydro-Meteorogical Forecasting [28], are input the program. For Mode 2, the weather data generation function in SOLSTILL does its job. The outputs of SOLSTILL consist of hourly amounts of distillate water, hourly temperatures of the cover, basin water and the basin, etc. In this study, only hourly amounts of distillate water are considered.

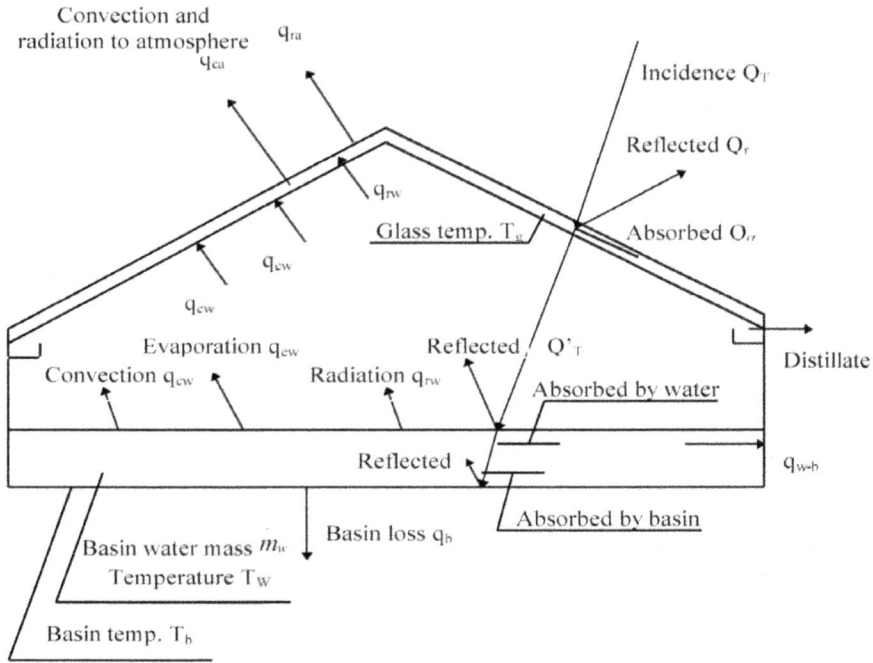

Figure 13.
The heat and mass transfer processes in a conventional solar still.

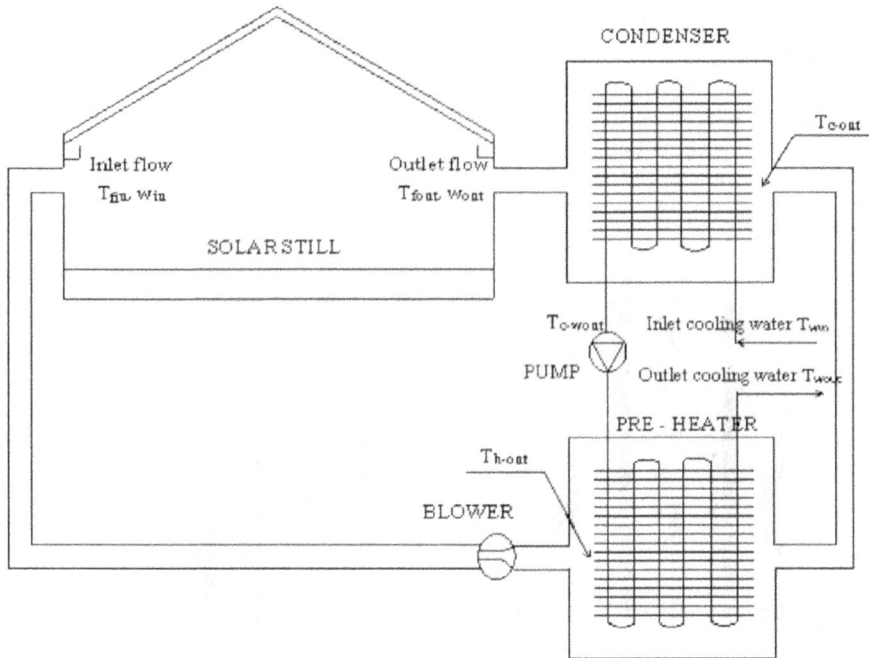

Figure 14.
Schematic diagram of a forced circulation solar still with enhanced water recovery.

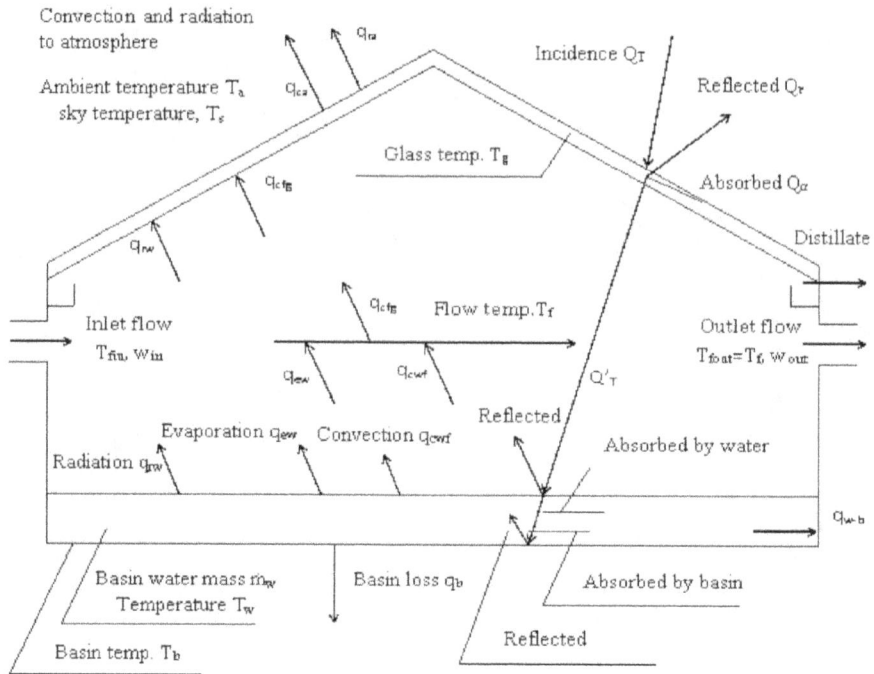

Figure 15.
The heat and mass transfer process in a forced circulation solar still.

Then, daily and monthly average daily amounts of distillate water are achieved. **Figures 16** and **17** show the monthly average daily distilled water of a conventional solar still for Ho Chi Minh City and Da Nang whereas **Figures 18** and **19** show those of a forced circulation solar still with enhanced water recovery, respectively.

As shown in **Figures 16–19**, the errors of the predicted monthly average daily distillate productivity of both a conventional passive solar still and a forced circulation solar still with measured and generated weather series as input data are very

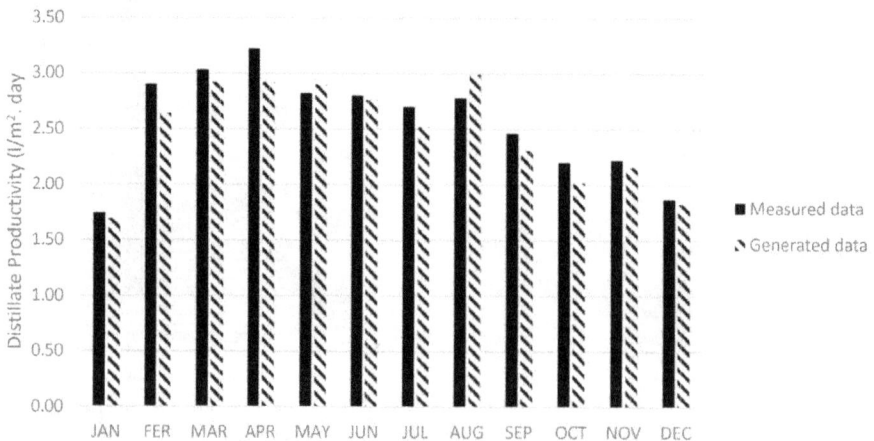

Figure 16.
Monthly average daily distillate productivity of a conventional solar still in Ho Chi Minh City.

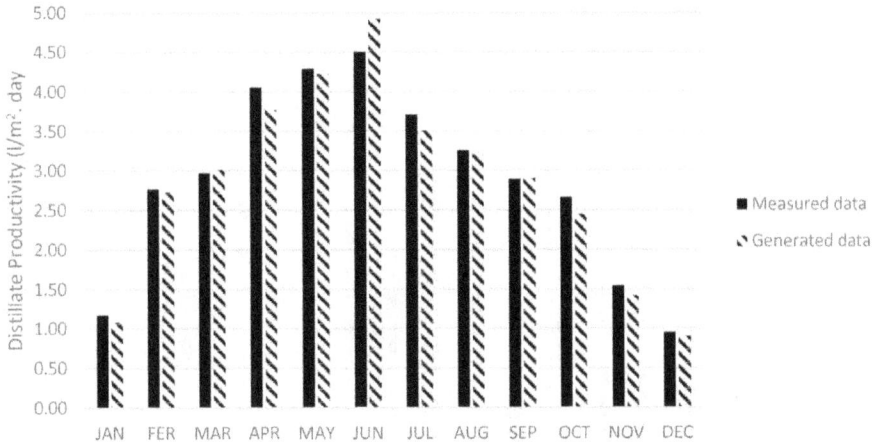

Figure 17.
Monthly average daily distillate productivity of a conventional solar still in Da Nang.

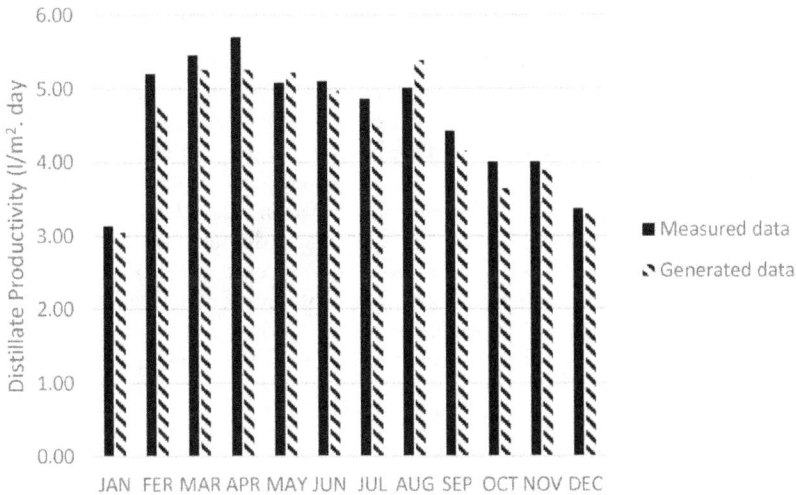

Figure 18.
Monthly average daily distillate productivity of a forced circulation solar still in Ho Chi Minh City.

small. The largest error is 9.3%, occurred in April in Ho Chi Minh City for a conventional solar still. The errors of predicted yearly average daily distillate productivities are less than 5%. Therefore, it can be expected that the weather data generated from the proposed models can be used to run any simulation programs for solar distillation systems.

6. Conclusion

In this study, to generate daily clearness index sequences for Ho Chi Minh City and Da Nang, two cities presenting for two climate types in tropical region, Aguiar's model was chosen. Then a modified model of Graham was proposed to generate hourly clearness index sequences from generate daily clearness index series for these

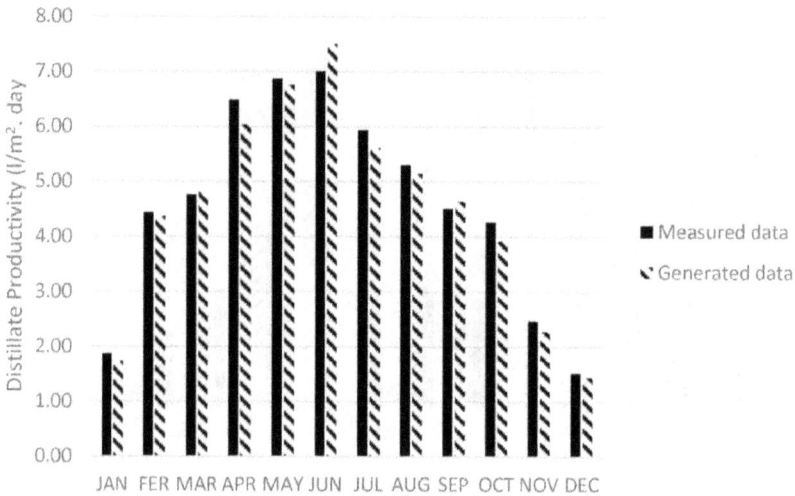

Figure 19.
Monthly average daily distillate productivity of a forced circulation solar still in Da Nang.

two locations. After that, a model to generate hourly ambient temperature sequences from monthly average daily ambient temperatures was presented. Having been proved by some statistic configurations and the predicted distillate productivities of solar still simulations, the models in this study are accurate in predicting daily and hourly irradiances and ambient temperature sequences. Especially, the model proposed in this study to generate the hourly solar radiation values is much simpler compared with Graham model. Therefore, both solar radiation and ambient temperature generating models in this work are believed to be used to calculate daily and hourly weather data for any numerical simulation programs of solar distillation systems with very limited input parameters, including the latitude, monthly average daily clearness index and ambient temperature values of the investigated locations.

Author details

Bao The Nguyen[1,2]

1 Ho Chi Minh City University of Technology, Vietnam National University, Ho Chi Minh City, Vietnam

2 Institute of Sustainable Development – ISED, Ho Chi Minh, Vietnam

*Address all correspondence to: thebao@hcmut.edu.vn

IntechOpen

References

[1] J. Duffie, and W. Beckman, Solar Engineering for Thermal Processes - 4th Edition. John Wiley & Sons, Inc., Hoboken, New Jersey, 2013.

[2] J. Boland, "Time series modelling of solar radiation". Modelling Solar Radiation at the Earth's Surface, V. Badescu (Ed.), Springer-Verlag Berlin Heidelberg, 2008, pp. 283-312.

[3] B. Nguyen, and V. Hoang, "The study of direct solar radiation data in the project of mapping the solar resource and potential in Vietnam". The 5th International Conference on Sustainable Energy 2017, Ho Chi Minh City, Vietnam.

[4] C. Fernandez-Peruchena, L. Ramirez, I. Pagola, and M. Gaston, "Assessment of models for estimation of hourly irradiation series from monthly mean values". 15th SolarPACES Conference, Berlin, Germany, pp.121-126. Hal-00919043.

[5] K. Brecl and M. Topic, "Development of a stochastic hourly solar irradiation model". International Journal of Photoenergy Volume 2014, DOI: 10.1155/2014/376504, 7 pages.

[6] J. Bright, C. Smith, P. Taylor, and R. Crook, "Stochastic generation of synthetic minutely irradiance time series derived from mean hourly weather observation data". Solar Energy 115, pp. 229-242.

[7] M. Hofmann, S. Riechelmann, C. Crisosto, R. Mubarak, and G. Seckmeyer, "Improved synthesis of global irradiance with one-minute resolution for PV system simulations". International Journal of Photoenergy Volume 2014, DOI: 10.1155/2014/808509. 10 pages.

[8] T. Soubdhan, and R. Emilion, "Stochastic differential equation for modelling global solar radiation sequences". Modelling, Identification and Control, GOSIER, France. DOI: 10.2316/P.2010.702-099. Hal-01823269.

[9] L. Magnano, J. Boland, and R. Hyndman, "Generation of synthetic sequences of half hourly temperature". Environmetrics, Vol. 19, No. 8, pp. 818-835.

[10] L. Mora-Lopez, "A new procedure to generate solar radiation time series from machine learning theory". Modelling Solar Radiation at the Earth's Surface, V. Badescu (Ed.), Springer-Verlag Berlin Heidelberg, 2008, pp. 313-326.

[11] T. Gafurov, J. Usaola, and M. Prodanovic, "Incorporating spatial correlation into stochastic generation of solar radiation data". Solar Energy 115, pp. 74-84.

[12] W. Hammed, B. Sawadi, S. Al-Kamil, M. Al-Radhi, and R. Abd-Alhameed, "Prediction of solar irradiance based on artificial neural networks". Invention Vol. 4, No. 45. DOI: 10.3390/inventions40300045.

[13] M. Vakili, S. Sabbagh-Yazdi, and K. Kalhor, "Using artificial neural networks for prediction of global solar radiation in Tehran considering particulate matter air pollution". Energy Procedia. Vol. 74, pp. 1205-1212.

[14] S. Priya, and M. Idbal, "Solar radiation prediction using artificial neural networks". International Journal of Computer Applications Vol. 116, No. 6.

[15] F. Tymvios, S. Michaelides, and C. Skouteli, "Estimation of surface solar radiation with artificial neural networks". Modelling Solar Radiation at the Earth's Surface, V. Badescu (Ed.), Springer-Verlag Berlin Heidelberg, 2008, pp. 221-246.

[16] A. Sfetsos, and A. Coonick, "Univariate and multivariate forecasting of hourly solar radiation with artificial intelligence techniques". Solar Energy 68, pp. 169-178.

[17] J. Wu, and C. Chan, "Prediction of hourly solar radiation using a novel hybrid model of ARMA and TDNN". Solar Energy, vol. 85, pp. 808-817.

[18] L. Mora-Lopez, J. Mora, R. Morales-Bueno, and M. Sidrach-de-Cardona, "Modelling time series of climatic parameters with probabilistic finite automata". Environmental Modelling & Software, Vol. 20, pp. 753-760.

[19] R. Aguiar, M. Collares-Pereira, and J. Conde, "Simple procedure for generating sequences of daily radiation values using a library of Markov transition matrices". Solar Energy 40, pp, 269-279.

[20] V. Graham and K. Hollands, "A method to generate synthetic hourly solar radiation globally". Solar Energy, Vol. 44, No. 6, pp. 333-341.

[21] B. Nguyen, and T. Pryor, "Generating artificial weather date sequences for Australian conditions". The 34th Annual Conference of ANZSES – Solar'96: Energy for Life, Darwin, Australia, pp. 101-108.

[22] C. Mustacchi, V. Cena, and M. Rocchi, M. "Stochastic simulation of hourly global radiation sequences". Solar Energy, 23, pp. 47-51.

[23] R. Aguiar and M. Collares-Pereira, TAG: a time dependent-, autoregressive, Gaussian model for generating synthetic hourly radiation. Solar energy 49, pp. 167-174.

[24] HOMER Energy LLC. Hybrid Optimization of Multiple Energy Resources, Ver 3.14.0. HOMER Energy, Boulder, CO, USA, 2020.

[25] K.M.Knight, S.A. Klein and J.A. A methodology for the synthesis of hourly weather data. *Solar Energy* **46**, 1991, pp. 109-120.

[26] D.G.Erb, S.A. Klein and W.A. Beckman. Estimation of degree-days and ambient temperature bin data from monthly average temperatures. ASHRAE Journal **25**, 1983, pp. 60-65.

[27] B. Nguyen. The Mathematical Model of Basin-Type Solar Distillation Systems, Distillation - Modelling, Simulation and Optimization, Vilmar Steffen, IntechOpen, 2019, DOI: 10.5772/intechopen.83228. Available from: https://www.intechopen.com/books/distillation-modelling-simulation-and-optimization/the-mathematical-model-of-basin-type-solar-distillation-systems

[28] National Center for Hydro-Meteorogical Forecasting. http://www.nchmf.gov.vn

Chapter 4

Desalination by Membrane Distillation

Mustakeem Mustakeem, Sofiane Soukane,
Muhammad Saqib Nawaz and Noreddine Ghaffour

Abstract

At present, around 25% of water desalination processes are based on distillation. Similar to classical distillation, membrane distillation is a phased-change process in which a hydrophobic membrane separates two phases. Membrane distillation is considered an emerging player in the desalination, food processing and water treatment market. Due to its high salt rejection, less fouling propensity, operating at moderate temperature and pressure, membrane distillation is considered as a future sustainable desalination technology. The distillation process is quite well known in desalination. However, membrane distillation emerged a few decades ago, and a thorough understanding is needed to adapt this technique in the near future. This review chapter introduces the classical distillation and membrane distillation as an emerging technology in the desalination arena. Heat and mass transfer and thermodynamics in membrane distillation, characteristics of the performance metrics of membrane distillation are also described. Finally, the performance evaluation of MD is presented. The possibility of using low-grade heat in membrane distillation allows it to integrate directly to solar energy and industrial waste heat.

Keywords: membrane distillation, desalination, vapor flux, vapor transport, evaporation

1. Introduction

Distillation is a thermal process in which a component separates out from a multi-component mixture by a phase-change process. When a membrane is put in between the feed and condensing solution, the feed can be vaporized and condensed at the membrane interface (thickness $\approx 200 \ \mu m$). **Figure 1** shows a schematic of a single-stage distillation and membrane distillation setup. The membrane distillation (MD) process combines the use of conventional distillation and membranes processes. It is a hybrid technology that uses the advantages of membrane separation and thermal distillation processes. In MD, the separation process employs a porous hydrophobic membrane between feed and permeate, allowing only solvent vapor to pass through, retaining the liquid/solid phase. Although the membrane provides a mass transfer resistance to the vapors, its employment allows water to condense within a minimal distance. This gives an advantage in creating a large partial pressure gradient across the membrane.

In a typical MD setup, the trans-membrane temperature difference (temperature difference between two sides of the membrane) creates the vapor pressure

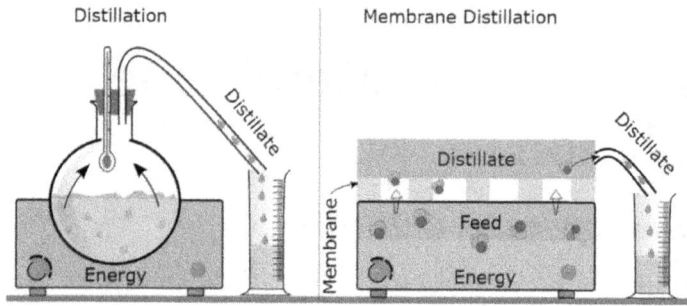

Figure 1.
A schematic of a typical single stage distillation system and the membrane distillation system. In MD, the vapors generate at the feed-membrane interface, and move to the permeate side and membrane allows only vapors to pass through.

difference, which drives mass transport [1–5]. The vapor is generated at the feed side and moves through the membrane pores to condense/get collected at the permeate side of the membrane. The salts, due to their non-volatile nature, remain in the feed solution.

MD has advantages over conventional desalination techniques. These include [6]:

- Low feed temperature in the range 50–90°C.

- Low fouling propensity.

- Capable to treat high saline water.

- High product water quality.

- Good compromise between specific enthalpy and efficiency.

Based on the distillate collection methods, MD can be classified into four broad configurations: (1) air gap membrane distillation (AGMD); (2) sweeping gas membrane distillation (SGMD); (3) vacuum membrane distillation (VMD); and (4) direct contact membrane distillation (DCMD) [7]. In all configurations, the feed solution remains in direct contact with one side of the membrane, while on the other side, distillate collection differs based on the type of the variant.

1. *Air gap membrane distillation*: In AGMD, the permeate side has an air gap followed by a condensing plate. A typical AGMD set up is shown in **Figure 2 (A)**. The vapor transfers from the hot feed to the air gap and eventually condenses at the cooling plate. The cooling plate can be cooled with multiple ways such as liquid cooled, air cooled, evaporative cooled; and gives the flexibility to use any coolant liquid as it does not mix with the condensate. AGMD is the most practical configuration in water production. However, due to high mass transfer resistance at permeate side, the flux is less as compared with other configurations. The air gap reduces the conduction heat loss from feed to permeate [8–10].

2. *Sweeping gas membrane distillation*: The SGMD uses a stream of gas to strip off the vapors from the permeate channel, after which they are either condensed externally or discarded as waste. **Figure 2(B)** shows a typical SGMD set up. SGMD shows less conduction heat loss than DCMD due to less conductivity of

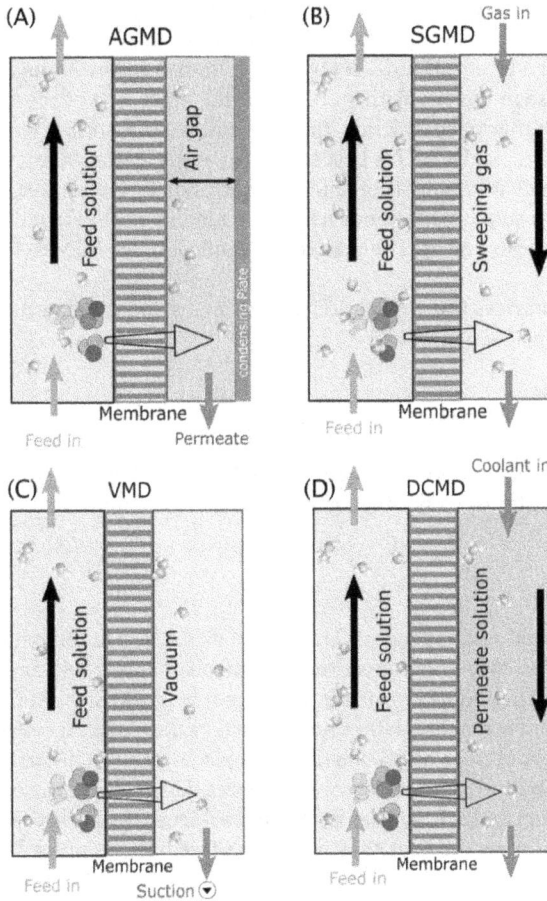

Figure 2.
A schematic of the four different MD configuration (A–D). The feed water flows tangentially in a cross flow regime at the feed side. At the permeate side, distillate taken off through various mechanisms and membrane allows only vapors to pass through.

gases at the permeate side. The pore wetting possibility is very less as the flow of gas takes off the distillate without condensing it [11]. The main disadvantage of this configuration is the high gas flow to sweep a unit volume of the permeate product. SGMD can be used in desalination and concentration of non-volatile liquids [7, 8].

3. *Vacuum membrane distillation*: The VMD and SGMD are very much similar in stripping out the permeate product. The VMD uses vacuum suction to strip off the vapors from permeate channel and either condense those externally or discarding as waste. **Figure 2(C)** shows a typical VMD configuration. The main advantage of this configuration is to separate the volatile organics from the main stream. However, it requires an additional pump and a condenser, increasing both CAPEX and OPEX. Also, the possibility of wetting the membrane is high due to the negative pressure at the distillate side [5, 8].

4. *Direct contact membrane distillation*: The DCMD is the most widely used MD configurations due to its simplicity, low capital cost and comparatively higher

flux [12–14]. In DCMD, as the name suggests, the membrane remains in direct contact with both solutions, i.e., feed and permeate. **Figure 2(D)** shows a schematic of a typical DCMD set up. The vapors transfer through the membrane and condense in the circulating coolant at the other side of the membrane. To achieve high thermal efficiency, the feed and coolant stream runs in counter-current directions. DCMD is best suited for applications where water is a major permeate, such as desalination and concentration of fruit juices. Also, the condensation step is carried out inside the module itself. High conduction losses across the membrane are the main disadvantage of DCMD [8, 15].

Table 1 summarizes the applications, advantages and disadvantages of different MD configurations [4, 6–9, 11, 15–17].

1.1 Transport phenomena

The temperature gradient between the feed and distillate sides of the MD system result in heat transfer from hot feed to cold distillate accompanied by mass transfer.

1.1.1 Heat transfer

MD is a non-isothermal process. Due to the difference in temperature at feed and permeate sides, three main heat transfer mechanisms take place: convective heat transfer, conduction heat transfer, and latent heat transfer. At the feed side and permeate side, convective and latent heat transfer take place. Across the membrane, conductive and convective heat transfer takes place. The two interfaces show convection with the bulk fluid, and membrane pores demonstrate the conduction phenomenon associated with vapor heat transfer. Due to a positive temperature gradient, it is clear that the heat transfer takes place only from feed to permeate.

MD configuration	Application area	Advantage	Disadvantage
DCMD	• Seawater desalination • Industrial wastewater • Dye effluents	• High distillate flux • Simple design and operation	• High conductive heat loss • High heat transfer coefficient
AGMD	• Seawater desalination • Industrial wastewater	• Low conductive loss • Low TP • Less chances of distillate contamination • High thermal efficiency	• Low permeate flux • High mass transfer resistance
SGMD	• Brackish water desalination • Azeotropic mixture separation • VOC removal	• Low conductive loss • Less pore wettability • No in-system condensation	• High gas volume needed
VMD	• Seawater desalination • Aroma recovery • Industrial effluents • VOC removal	• Low conductive loss • High permeate flux • Condensate obtained externally	• Membrane wetting • High pump power • Heat recovery is difficult

Table 1.
Summary of the applications, advantages and disadvantages of different MD configurations [4, 6–9, 11, 15–17].

1.1.1.1 Feed side

At feed side, hot bulk solution come in contact with the MD membrane. Convective heat transfer Q_f (W·m^{-2}) takes place and can be expressed as per Newton's law of cooling [14]:

$$Q_f = h_f * (T_{fb} - T_{f,m}) \tag{1}$$

Where h_f is the feed convective heat transfer coefficient (W·m^{-2}K^{-1}), T_{fb} is the bulk feed solution temperature (K), $T_{f,m}$ is the temperature (K) at feed-membrane interface.

1.1.1.2 Membrane pore

Heat transfer through membrane occurs in two parallel routes: first is latent heat transfer Q_v at pores mouth and second is conduction heat transfer Q_c. The Q_c is the conduction heat loss and does not contributes to the vapor mass. Hence, for an efficient system, it is required to minimize Q_c as low as possible. It can be expressed by the Fourier's law of conduction. Therefore, thicker membrane contributes less to conduction heat transfer. The Q_m (W·m^{-2}) is the sum of both heat transfers and can be expressed as [2, 5][1]:

$$Q_m = Q_c + Q_v = \frac{k_m}{\delta_m} * (T_{f,m} - T_{p,m}) + J * h_{fg} \tag{2}$$

$$h_m = \frac{k_m}{\delta_m} \tag{3}$$

Where Q_c is the conduction heat transfer (W·m^{-2}), k_m is the membrane thermal conductivity (W·m^{-1}·K^{-1}), δ_m is the membrane thickness (m), J is the distillate flux (kg·m^{-2}h^{-1}), h_{fg} is the latent heat of evaporation (kJ·kg^{-1}), $T_{p,m}$ and $T_{f,m}$ are the temperature of permeate-membrane interface and feed-membrane interface respectively.

The thermal conductivity k_m (W·m^{-1}K^{-1}) of the membrane can be calculated from the thermal conductivity of vapor k_g and the polymer material k_p as per following relations [18]:

$$k_m = \varepsilon k_g + (1 - \varepsilon)k_p \tag{4}$$

Where ε is the porosity of the membrane.

1.1.1.3 Permeate side

Vapor that pass through membrane pores condense at the permeate side at the expense of latent heat of condensation. The heat transfer at the permeate side Q_p (W·m^{-2}) can be expressed as [15, 18]:

$$Q_p = h_p * (T_{pb} - T_{p,m}) \tag{5}$$

---

[1] The influence of mass transfer on heat transfer was ignored.

Where h_p is the permeate convective heat transfer coefficient (W·m^{-2}K^{-1}), T_{pb} is the bulk permeate solution temperature, $T_{p,m}$ is the temperature (K) at permeate-membrane interface.

1.1.1.4 Overall heat transfer

At steady state, heat transfer from feed to membrane, across the membrane and from membrane to permeate are equal. **Figure 3** shows with the electrical analogy of local heat transfer coefficients and their relationship with the overall heat transfer coefficient. The overall heat transfer Q (W·m^{-2}) can be expressed in terms of universal (overall) heat transfer coefficient as U [2, 18]

$$Q = Q_f = Q_m = Q_p$$

$$= h_f * (T_{fb} - T_{f,m}) = \frac{k_m}{\delta_m} * (T_{f,m} - T_{p,m}) + J * h_{fg} = h_p * (T_{pb} - T_{p,m}) \quad (6)$$

$$= U * (T_{f,m} - T_{p,m})$$

Where U (W·m^{-2}) can be expressed as:

$$\frac{1}{U} = \frac{1}{h_f} + \frac{1}{\frac{k_m}{\delta_m} * \frac{Jh_{fg}}{(T_{f,m}-T_{p,m})}} + \frac{1}{h_p} \quad (7)$$

The membrane interface temperatures, i.e. $T_{f,m}$ and $T_{p,m}$ can not be measured experimentally. Therefore, mathematical iterative procedure is generally used to evaluate both interface temperatures. The membrane interface temperature can be derived from the Eq. (6) and expressed as [15, 19]:

$$T_{f,m} = T_{fb} - \frac{J \, h_{fg} + \frac{k_m}{\delta_m} (T_{f,m} - T_{p,m})}{h_f} \quad (8)$$

$$T_{p,m} = T_{pb} - \frac{J \, h_{fg} + \frac{k_m}{\delta_m} (T_{f,m} - T_{p,m})}{h_p} \quad (9)$$

The heat transfer coefficients h_f and h_p can be calculated using Nusselt number relation as shown in Eq. (10), while the h_m can be calculated from Eq. (3). The Nusselt number correlations are available in the literature [2, 4, 5].

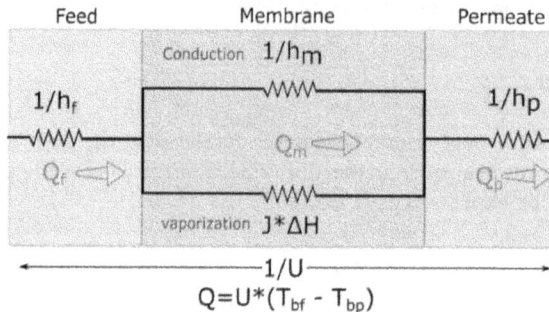

Figure 3.
Electrical equivalent of heat transfer resistance in MD process. An overall heat transfer resistance is an equivalent of all three heat transfer resistance of feed, membrane and permeate.

$$\text{Heat transfer coefficient} = \frac{Nu * k_m}{D_p} \qquad (10)$$

Where h_m is defined in Eq. (3)

1.1.2 Mass transfer

Mass transport in MD can be explained by three sequential stages: vapor generation, vapor transport, and vapor condensation, which respectively take place at the feed-membrane interface, membrane pores, and permeate-membrane interface. **Figure 4** shows the mass transfer process, including vapor generation at the feed side, vapor transport through the porous membrane, and finally, vapor condensation at the permeate side. There is resistance to mass transfer at each stage, which is defined by a correlation explained in the sequel. The overall mass transfer in the MD system is often expressed using Darcy's law i.e., the vapor pressure difference between two sides of the membrane, and is given as [1, 2]:

$$J = C_m * \left(p_{f,m} - p_{p,m} \right) \qquad (11)$$

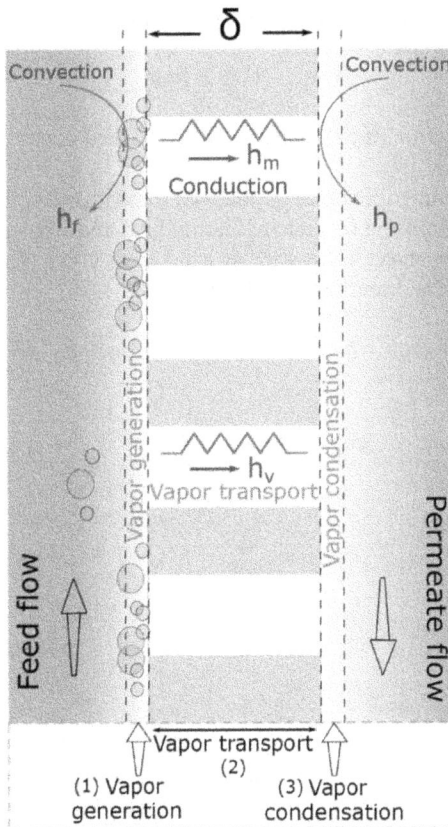

Figure 4.
A schematic of vapor transfer from vapor generation at the feed side, to the permeate side through the membrane pores. Mass transfer is the transport mechanism within the membrane pores. Finally, the distillate condenses at the permeate side.

Where C_m is the membrane mass transfer coefficient ($kg \cdot m^{-2} h^{-2} \cdot Pa^{-2}$), p_f and p_p are the partial pressures of water at feed and permeate sides, respectively. The three stages of mass transport are explained as below:

1.1.2.1 Stage 1 (Vapor generation)

At the feed side of the membrane, liquid feed remains in contact with the membrane pores. The vapors are generated at the feed-membrane pore interface when the hot feed solution passes over the hydrophobic MD membrane. The partial pressure of the vapor generated is directly proportional to the temperature as per Antoine's equation [19]. In the case of water, Antoine equation can be rewritten as Eq. (12):

$$p_v = exp \left[23.19 - \left(\frac{3816.44}{T_{f,m} - 46.13} \right) \right]$$ (12)

where p_v is the vapor pressure of water (Pa) and $T_{f,m}$ is the temperature at the feed-membrane interface (K).

The presence of non-volatile solute in the feed water decreases the vapor pressure of the feed solution. Therefore, the vapor pressure is a function of mole fraction of that component and can be expressed as [4]:

$$p_{(x)} = (1 - x_s) * p_w^o$$ (13)

Where x_s mole fraction of non-volatile solute in feed water and p_w^o is the partial pressure of pure water.

Figure 5 shows a control volume of heat and mass transfer of length dz. A feed of mass flow rate $\dot{m}$ fed into the feed channel, and heat and mass transfer takes place along the membrane surface. The width dy is taken as unity. The mass balance at feed side can be explained mathematically as [13, 20]:

$$\dot{m}_{f,(z+dz)} \; h_{fb} = \dot{m}_{f,(z)} \; h_{fb} - (J \; h_{fg} + q_m) dA$$ (14)

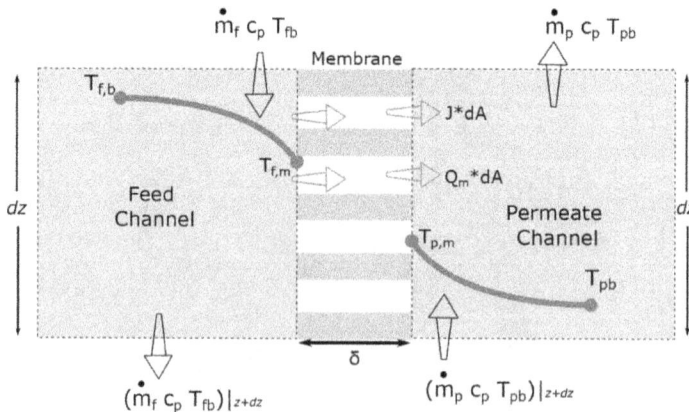

Figure 5.
A schematic of the mass balance in a control volume at the feed side and permeate side. The mass transfer takes place through the porous membrane along the vapor pressure gradient.

The mass flux of water that passes through the membrane can be expressed as:

$$J = k_f \, \rho_f \, \ln \frac{x_{f,m}}{x_f} \tag{15}$$

Where k_f is the mass transfer coefficient and can be calculated using Sherwood number as follow [4]:

$$Sh = \frac{k_f \, d_h}{D_s} \tag{16}$$

Where d_h is the hydraulic diameter (m), D_s is the solute diffusion coefficient in bulk feed (m$^2 \cdot$s^{-1}). The Sherwood number can be calculated using following semi-empirical relationship [4]:

$$Sh = \alpha \, Re^\beta \, Sc^\gamma \tag{17}$$

Where α, β, and γ are the coefficients calculated from experiments. Sc is the Schmidt number and Re is the Reynolds number, which can be expressed as [2]:

$$Sc = \frac{\eta_f}{\rho_f \, D_s} \tag{18}$$

$$Re = \frac{\rho_f \, v_f \, d_h}{\eta_f} \tag{19}$$

Where η_f is the viscosity of the bulk fluid (Pa·s), ρ_f is the density of the bulk fluid (kg·m^{-3}), and v_f is the fluid flow velocity (m·s^{-1}). Different correlations used for Sherwood number are listed in Appendix 1A.

1.1.2.2 Stage 2 (Vapor transport)

Two major factors control the vapor transfer in MD membrane pores. One is the vapor pressure difference Δp, and the second is the mass transfer coefficient of the membrane. The vapor transfer through the membrane may be the limiting step for mass transfer in MD, which is influenced by the physical properties of the membrane and other characteristics which expresses with the relation $J \propto \frac{D_p}{\chi} \frac{\varepsilon}{\delta_m}$ [1].

1. *Porosity* (ε): In general, regardless of the type of MD configuration, membranes with high porosity have more distillate flux as well as lower conductive heat loss. The porosity of MD membranes lies between 30 and 85% [4, 21].

2. *Tortuosity* (χ): In a simple assumption, membrane pores are considered as straight cylindrical channels. However, they possess many curved paths. High tortuosity value leads to lower distillate flux due to vapor permeation through tortuous paths. Therefore, membrane permeability is inversely proportional to the membrane tortuosity. In most theoretical models in MD studies, a tortuosity value of 2 is frequently considered to predict the transmembrane flux [15].

3. *Pore size*: Large pore size allows more vapor to pass through. However, after a certain point, it will limit the applied pressure to avoid pore wetting. On the other hand, small pore size enables working at high pressures but at the cost of lower fluxes [4].

The vapor transport mechanism through MD membrane pores is governed by three basic mechanisms known as Knudsen diffusion (K_n), Molecular-diffusion, and viscous (poiseuille flow). The mass transport through the membrane is described by the simple Darcy's law, expressed in Eq. (11). For dilute solutions, it can be written as [7, 22]:

$$J = C_m * \frac{dP}{dT}(T_{f,m} - T_{p,m}) \tag{20}$$

The Pressure term $\frac{dP}{dT}$ can be calculated from Clausius-Clapeyron equation as:

$$\frac{dP}{dT} = \frac{\Delta H}{RT^2} P_{av}(T_m) \tag{21}$$

The pressure P_{av} for non-ideal aqueous solution can be calculated from Raoult's law as expressed by Eq. (13).

As per Darcy's law Eq. (11), membrane mass flux is mainly governed by the partial pressure difference between feed and permeate side and the membrane mass transfer coefficient C_m. The membrane mass transfer coefficient is primarily a function of membrane properties (thickness, pore size, tortuosity) and the process conditions (temperature and pressure). Its value depends upon the mass transfer mechanism inside the membrane pore.

The air molecules (particles) act as a medium in the pores. Knudsen number (K_n) determines the type of governing mechanism of mass transport inside the membrane pore and can be expressed as [1]:

$$K_n = \frac{\lambda}{D_p} \tag{22}$$

Where D_p is the pore diameter (m) λ is the mean free path (m) of the vapor molecule and can be written as [19]:

$$\lambda = \frac{k_B T_r}{\sqrt{2}\pi P_m \sigma_v^2} \tag{23}$$

where k_B is the Boltzmann constant (1.38×10^{-23}·J·K^{-1}), P_m is the mean average pressure in membrane pores (Pa), T_r is the average temperature in the pore, and σ_v is the water vapor collision diameter (0.2641 nm).

Depending on the value of K_n, three possible mass transfer modes exist as follows: (a) Knudsen diffusion ($K_n > 1$) in which molecular collisions with the walls dominate as compared to the molecule-molecule collisions, (b) molecular diffusion ($K_n < 0.01$) in which the frequency of gas molecule collisions is much higher than those with the pore walls, and (c) Knudsen-molecular diffusion ($0.01 < K_n < 1$) in which the frequency of molecular collisions with the pore walls is similar to that of the gas-gas collisions (often referred to as "transitional regime") [4, 19]. **Figure 6** shows a schematic of three possible mass transfer mechanisms through the pores of an MD membrane.

    i. *Knudsen diffusion:* If the mean free path of water molecules is greater than the pore diameter ($\lambda > > D_p$), the frequency of collision between water molecules are less than the molecule–pore wall. The Knudsen number in this regime is ≥ 1 [19]. Therefore, it is the dominating process of mass transport in DCMD. The mass transfer resistance arises from the momentum transfer of vapor molecules with the sidewalls of the pores.

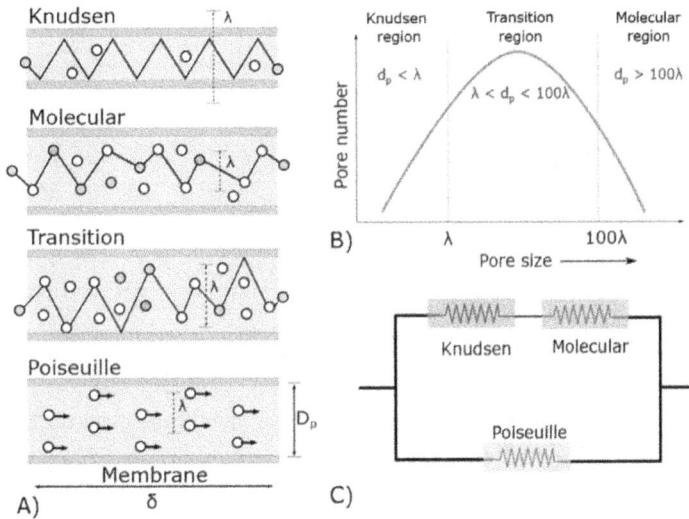

Figure 6.
(A) Various mechanisms of vapor transport through membrane pores are shown. (B) Mass transport regime based on the pore size. The transition regime occurs in the intermediate values of the Knudsen number. (C) Mass transport resistance and their electrical circuit analogy.

Therefore, Knudsen diffusion decreases with the increase of temperature. In this case, the membrane mass transfer coefficient is shown in **Table 2**.

ii. *Molecular diffusion:* Molecular diffusion occurs when the mean free path of vapor molecule ($\lambda < < D_p$) is much shorter than the pore diameter. The Knudsen number in this case is below 0.01 [5, 7]. Therefore, the intramolecular (within own molecules) and intermolecular (with the air molecules) collisions prevail vapor molecule-wall collisions. The mass transfer resistance comes from the collision of vapor molecules with the air molecules entrapped inside the membrane pores. In this case, the membrane mass transfer coefficient and mass transfer mechanism are shown in **Table 2** and **Figure 6** respectively. Since the feed and permeate in DCMD is deaerated, therefore, molecular diffusion in DCMD is minimum [2]. Similarly, in the case of VMD, the vacuum in the permeate side remove the entrapped air in the pore. Therefore, molecular diffusion is neglected there as well.

iii. *Poiseuille flow:* In poiseuille flow (viscous flow), vapor molecule acts as a continuous fluid flow under a pressure gradient between two sides of membrane. It occurs in deaerated membrane pores under a stream vacuum. It applies when the mean free molecular path of vapor molecule is smaller that the pore size ($d_p > 100\lambda$) [2, 25]. The membrane mass transfer resistance is due to momentum transfer to the pore walls through viscous drag and its coefficient is shown in **Table 2**.

iv. *Knudsen-molecular diffusion:* Transport of water vapors is a complex phenomenon and can encounter mass transfer resistance from both air molecule and the pore walls. In practical, there exist more than one type of transport phenomenon, therefore, called transition flow. This type of transport occurs when the mean free path of the vapor molecule is in the range (1–100) λ and the flow is a kind of transition flow between pure

Knudsen number	Type of transport	Condition	Mass transfer coefficient C_m
>1	Knudsen diffusion	$\lambda >> D_p$	$\frac{2}{3}\frac{\varepsilon}{\chi}\frac{r}{\delta}\sqrt{\frac{8\,M_w}{\pi\,R\,T}}$
$0.01-1$	Knudsen-molecular diffusion	$\lambda < D_p < 100\lambda$	$\left[\left(\frac{\chi\,\delta}{\varepsilon}*\frac{R\,T_m}{M_w}\frac{P_a}{P_T}\frac{1}{D_w}\right)+\left(\frac{(3\,\chi\,\delta)}{(2\,\varepsilon\,r)}\sqrt{\frac{\pi\,R\,T}{8\,M_w}}\right)\right]^{-1}$
<0.01	Molecular diffusion	$\lambda << D_p$	$\left(\frac{\varepsilon}{\chi\,\delta}\frac{M_w}{R\,T_m}\frac{P_T}{P_a}(D_w)\right)$
<0.01	Poiseuille flow	$\lambda << D_p$	$\left(\frac{1}{8}\frac{r^2}{\chi}\frac{\varepsilon}{\delta}\frac{M_w}{\mu}\frac{1}{R\,T_m}P_m\right)$

Table 2.
Mass transfer mechanism within the membrane pores follow a specific regime based on the Knudsen number. The mass transfer coefficient (C_m) of each mechanism is shown [1, 2, 4, 5, 19, 23, 24].

Knudsen, and molecular diffusion [5]. The membrane mass transfer coefficient is shown in **Table 2**.

1.1.2.3 Stage 3 (Vapor condensation)

At permeate side, each configuration of MD uses different methods to collect permeate. In some cases, the permeate is the waste and is discarded, while in other instances, permeate is the product and collected using various methods based on the configuration. In DCMD, the permeate is stripped by circulating pure water. Therefore, it offers no mass transfer resistance. In the case of VMD, the presence of vacuum also negates the existence of mass transport resistance.

However, AGMD show a significant resistance due to the presence of an air gap between the membrane and the condensing plate. In AGMD, the mass transfer occurs through molecular diffusion between the membrane pore surface to the condenser plate. The mass transfer resistance can be expressed by:

$$k_s = \frac{P_{av}}{P_a}\frac{\varepsilon}{\delta\,\gamma+b}\frac{D_w\,M_w}{R\,T_m}$$ (24)

Where b is the thickness of the air gap (m) [2, 4].

1.1.3 Temperature polarization

The phenomenon of evaporation at the feed-membrane interface and condensation at the permeate-membrane interface creates a temperature gradient with the bulk solution. Temperature polarization (TP) is a condition when the temperature at the membrane interface differs from its bulk solution [26–28]. TP is considered a critical factor that impacts the vapor flux of an MD system. The evaporation phenomenon at the liquid-air interface draws the latent heat from the bulk solution. Similarly, at the permeate side, the liquid-membrane interface releases heat of condensation to the coolant liquid. This creates a temperature difference between the bulk solution and the membrane interface. **Figure 7** shows a schematic representation of temperature profile across the membrane [28].

There exist a thermal boundary layer at each side of the membrane. However, this thermal boundary layer does not have a significant effect in the case of AGMD, VMD, and SGMD. Moreover, the salt concentration at the feed-membrane interface increases due to mass transfer, leading to the concentration polarization (CP) phenomenon. However, the effect of CP is negligible in MD [18, 29]. The operation parameters such as fluid velocity, concentration, and temperature of feed solution

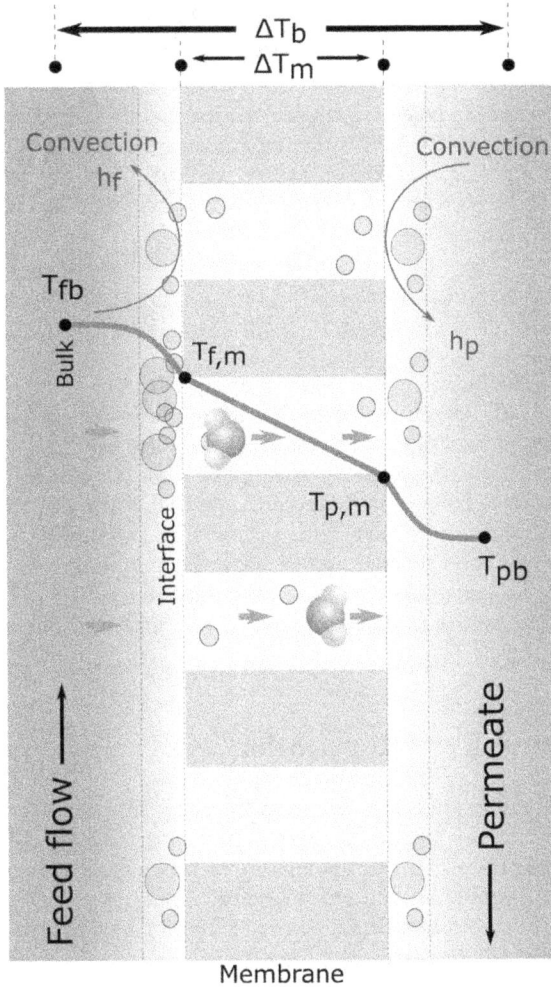

Figure 7.
A profile of temperature deviation at interfaces with the bulk fluid on both sides of the membrane. The temperature change at the membrane interface ΔT_m and bulk ΔT_b is shown to determine the polarization coefficient.

affect the TP. Higher velocity increases the heat transfer by creating turbulences locally, hence diminishing the thermal boundary layer. Additionally, the lower concentration and higher temperature of the feed solutions produce high vapor pressure as per Rault's law and Antoine's Eq. (12) respectively. All the above parameters decrease the effect of TP. In addition, the membrane also plays an important role in TP due to the heat transfer across it. High porosity decreases the TP, while higher thermally conductive polymers show high TP. However, high thickness decreases the heat loss across the membrane as per Fourier's law Eq. (2) and hence decreases TP. It is observed that TP decreases the vapor flux significantly [21].

1.1.3.1 Temperature polarization coefficient

The quantification of TP is expressed in temperature polarization coefficient (TPC). **Figure 7** shows the temperature difference at the membrane interfaces with its bulk. TPC represented as Θ is expressed as [19]:

$$\Theta = \frac{\Delta T_m}{\Delta T_b} = \frac{\left(T_{f,m} - T_{p,m}\right)}{\left(T_{fb} - T_{pb}\right)} \tag{25}$$

The thermal boundary layer at both sides of the membrane acts as resistances to heat transfer. In other words, TPC is the ratio of thermal boundary layer resistance to the total heat transfer resistance. In the case of VMD, TPC can be simplified as the ratio of feed-membrane interface temperature to the bulk feed temperature [1], as expressed in Eq. (26).

$$\Theta = \frac{T_{f,m}}{T_{fb}} \tag{26}$$

The TPC value, lies between 0–1 and most of the literature reports results between 0.4 and 0.7 [1, 3, 5, 30]. When the TPC value approaches zero, the system is limited by heat transfer at the feed side, which indicates an inefficient design. In contrast, TPC value 1 indicates that the system is affected by mass transfer resistance.

Several studies have been proposed to mitigate TP by using turbulence promotors such as spacers. However, this approach creates additional energy requirements [27, 31–34]. Considering that TP and conduction heat loss is an intrinsic process deficiency that cannot be fully mitigated, it is highly desirable to seek alternative approaches to alleviate heat loss and achieve a sustainable MD performance.

2. Parameters that affect MD vapor flux

Most of the MD research is focused on maximizing vapor flux. However, taking vapor flux as a matrix to evaluate the thermal performance, may not be a correct approach since vapor flux depends upon many factors including MD system configurations, active membrane area, type of energy input, heat recovery from exiting feed etc. Therefore, the highest flux may not lead to the best thermal efficiency. Following are the parameters affecting the vapor flux:

1. *Effect of feed temperature*: The feed temperature is the most important factor and has a major effect on the vapor flux of an MD system. An exponential increase in vapor flux was observed with the increase in inlet feed temperature. This is due to the exponential increase of vapor pressure with temperature resulting in increase in driving force. **Figure 8** shows the effect of feed temperature on permeate vapor flux on DCMD.

   Further, the conduction heat transfer is also higher at higher feed temperatures, leading to higher temperature polarization. Though, at high inlet feed temperature, the thermal efficiency is higher due to higher vapor flux, as shown in the **Figure 8**.

2. *Effect of permeate temperature:* In most of lab scale MD systems, the permeate temperature varies from 10 to 40°C. Increase in permeate temperature resulted in decrease in trans-membrane driving force, hence, vapor flux decreases. The effect varies from linear to exponential depending upon the membrane module characteristics [15, 35].

3. *Effect of feed and permeate flow rate:* The feed and permeate velocity influence the temperature polarization by creating the hydraulic turbulences in the feed

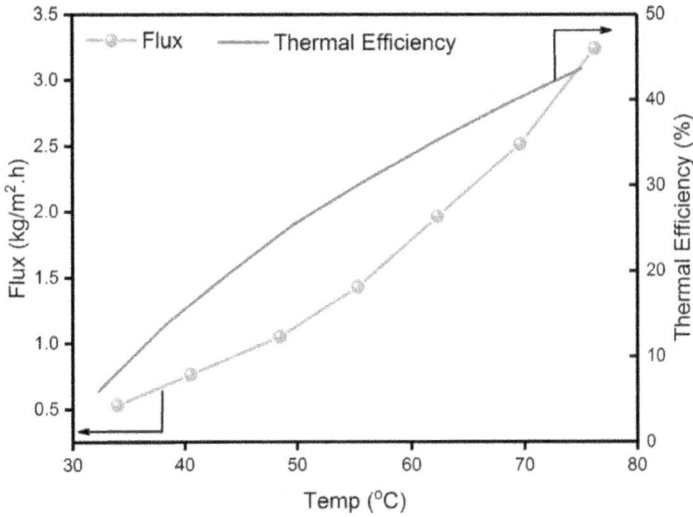

Figure 8.
Graph showing temperature effect on vapor flux of a DCMD configuration. The Line graph shows the variation in thermal efficiency with feed inlet temperature. Adapted from [15].

and permeate channel respectively. The turbulences increase the heat and mass transfer coefficient (h_f, h_p) at the MD membrane's feed and permeate boundary layer. The increased flow rate decreases the effect of concentration polarization. The vapor flux increases to an asymptotic value with an increase in feed velocity. Further, the thermal boundary layer thickness decreases at higher feed velocities. Kubota et al. (2020) revealed that the permeate flux increases with the feed velocity until it reaches a maximum and then decreases [7, 36].

4. *Effect of the presence of non-condensable gasses:* Non-condensable gasses such as CO_2 present in the feedwater stream. These gasses may mask the membrane pores, producing mass transfer resistance. Feedwater is generally aerated with N_2 gas before feeding in the MD system. However, this effect does not affect the VMD process. The process of deaeration consumes energy, and hence it decreases the overall performance [36].

5. *Effect of solute concentration:* High salt concentration decreases the vapor flux due to decrease in vapor pressure as per Roult's law. Additionally, concentration polarization may develop at the vicinity of the membrane, producing mass transfer resistance. A reduction in coefficient of polarization (θ) and thermal efficiency were observed with the increase in salt concentration [37]. Martinez et al. demonstrated that vapor flux decreased only by 1.15 fold by increasing the salt concentration by fivefold [38]. Therefore, MD can be used to treat high saline water with a slight decrease in productivity [1, 39].

6. *Effect of module geometry:* The configuration arrangement and the modular dimensions affect the TP in an MD system. Longer channel (keeping the width constant) leads to higher water production capacity, however, the flux decreases due to continuous decrease in feed water temperature along the length. The higher water production capacity is due to high water residence time with the MD membrane [40].

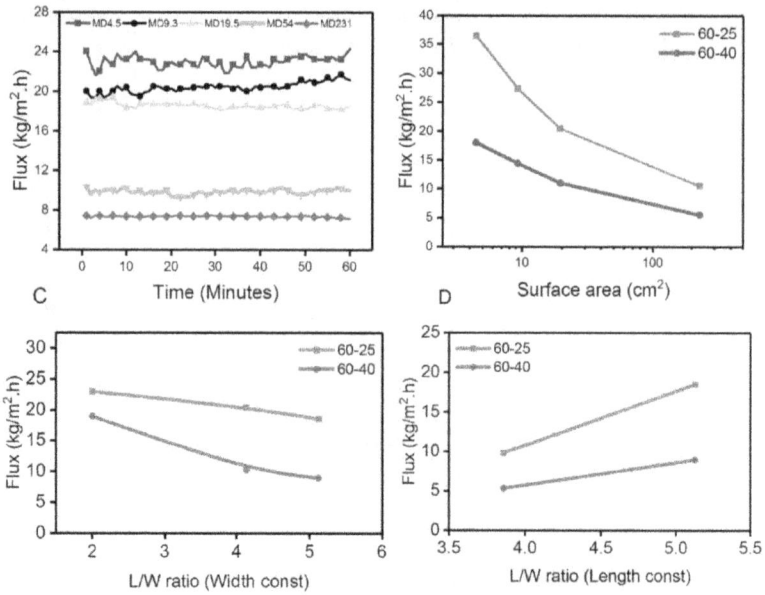

Figure 9.
Vapor flux variation with respect to the modular dimensions of various MD flow cell sizes.

As the feed water moves along the length (with constant width), it loses heat due to latent heat and heat of conduction, which results in temperature decrease. Therefore, the permeate flux decreases. However, due to more residence time, the total water capacity increases. Therefore, there is an optimum length of the feed channel length. The water flux variation with the length, length/width ratio, surface area have been investigated to underline the affect. [40] presented a mathematical model simulation which shows that the total residence time of feed water in the feed channel has a positive effect until a limit, after that it started decreasing. The peak of the curve shows the optimum length of the feed channel.

Figure 9 shows the vapor flux data of four MD module types having surface area 4.5, 9.3, 19.5, 54, and 231 cm^2. The vapor flux was plotted with various variables. **Figure 9(A)** It is observed that the small active surface area membranes performance is higher than large surface area. This was a general consideration because the MD types were of different length and width. This can be deduced that regardless of width and length, the flux of lower active membrane surface area is higher than the larger ones, which is shown in the **Figure 9(B)**. This shows a negative exponential correlation of the flux with the surface area.

The length has a negative effect on the vapor flux as demonstrated in a simulation by Lee et al. [40]. Our experimental data verifies this relation as shown in **Figure 9(C)**. If the length was kept constant, increasing the width was a positive effect on vapor flux [40]. The length to width ratio versus flux is plotted in the **Figure 9(D)** shows a positive slop.

As the length increases, the temperature of the feed drops due to the effect of latent heat and conduction heat loss along the length. The temperature drop increases. This was confirmed by the simulated results demonstrated by [40].

3. Membrane properties

Unlike Reverse Osmosis, MD membranes are not chemically involved in the mass transfer phenomenon. However, they are involved in the heat transfer

phenomenon. MD membranes require some specific physical and chemical charac-
teristics to perform well in the MD process. Mainly two types of membranes are
used: Polyvinylidene Fluoride (PVDF) and Polytetrafluoroethylene (PTFE).
Figure 10(A) shows the required characteristics of an ideal MD membrane, and
Figure 10(B–C) shows an SEM topography of typical PVDF and PTFE membranes.
PVDF membranes have round pores that could be seen on the top surface, while
PTFE has pores strangled between the polymer fibers. The required physical and
chemical characteristics of an MD membrane include:

1. *High Porosity (ε)*: Membrane porosity is the volume of pores to the total
   volume of the membrane. Membranes with high porosity produce more
   distillate flux due to more available channels for mass transfer and lower
   conductive heat loss. However, it decreases the mechanical strength and make
   membrane prone to crack under mild pressure. Porosity of MD membranes
   lies between 30 and 85% [21].

2. *Adequate thickness (δ$_m$)*: The membrane thickness offsets both membrane
   permeability and heat transfer. Thicker membranes are suitable to prevent
   heat loss but will affect water vapor permeability [21].

3. *Lower tortuosity (χ):* Tortuosity is the deviation of pores from a straight
   cylindrical path in porous media. In a simple assumption, membrane pores can
   be considered as straight cylindrical. However, this assumption is often far
   from reality. High tortuosity values lead to lower distillate flux due to vapor
   permeation through tortuous paths. Therefore, membrane permeability is in
   inverse relation to membrane tortuosity. In most theoretical models in MD
   studies, a value of 2 is frequently considered to predict the transmembrane
   flux [15].

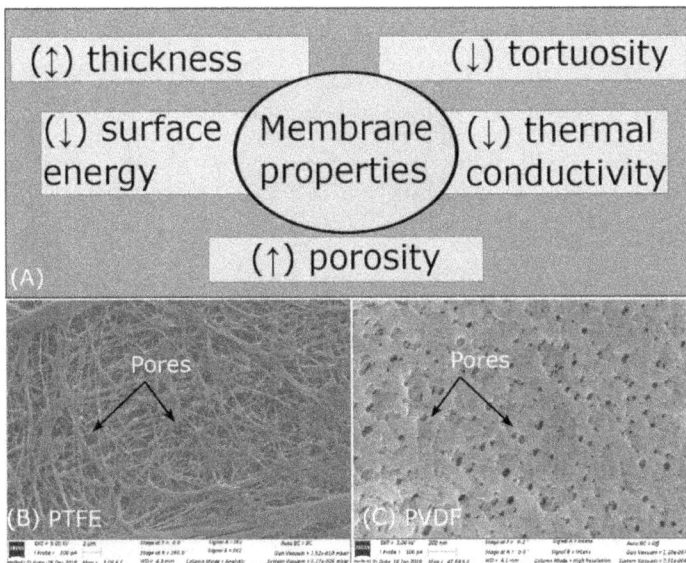

Figure 10.
*(A): The typical characteristics of an ideal MD membrane. (B–C): A surface topography of PTFE and PVDF
membranes through SEM imaging. The pores in a typical PVDF and PTFE membranes are also depicted.*

4. *Low thermal conductivity* (k_m): The high thermal conductivities facilitate sensible heat transfer, which leads to a decrease in temperature difference across the two sides of the membrane. Hence, vapor flux decreases. Therefore, low thermal conductivity is desired for MD membranes [22].

5. *Low surface energy:* Low surface energy correlates with high hydrophobicity. Materials with high hydrophobicity can transform to membranes with larger pores, which eventually increases the vapor flux. Also, high hydrophobic materials allow membrane processes to operate at high pressure for a given pore size [22].

3.1 MD module design

DCMD is one of the most used MD configurations due to its simplicity. Different MD modules have been developed to address a variety of process requirements. Based on membrane type and design principle, MD modules have been developed into three different designs:

1. Flat sheet module: In the flat sheet module, a flat sheet membrane is put tangentially with the feed flow. Flat sheet membranes offer simple fabrication and assembly, easy cleaning and maintenance. The main disadvantages of the flat sheet are its low packing density, and high propensity to wetting [41]. **Figure 11(A)** shows a typical set up of flat sheet DCMD.

2. Hollow fiber module: In this type of module, hollow fiber membranes are used. Membranes are set in parallel bundles in the shell of a cylindrical casing. The main advantage of the hollow fiber module is the high packing density (surface area to volume ratio) and lack of membrane support. However, the

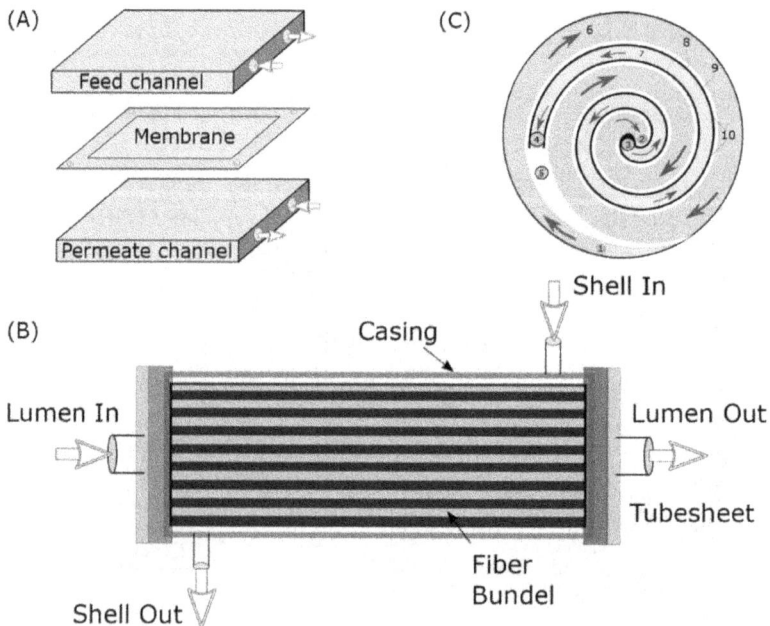

Figure 11.
(A): Flat sheet membrane module, showing a rectangular membrane between two channels. (B): A schematic of Hollow Fiber module set up, comprising hollow fiber membrane cased in a housing. (C): A spiral wound DCMD module commercialized by Solar Springs, adapted from [42].

permeate flux of the hollow fiber module is lower to that of the flat sheet. Additionally, the cleaning process and the membrane replacement is difficult in the hollow fiber module [4, 41]. **Figure 11(B)** shows a typical set up of hollow fiber DCMD.

3. Spiral wound: Flat sheets are spirally coiled in a cylindrically wound shape. The membrane is rolled with the spacer between the feed and coolant fluids, creating evaporator and condenser channels. Such type of system was first commercialized by Solar Springs as Oryx unit as shown in **Figure 11(C)** [42]. The condensation process first preheats the cold feed as a condenser stream in the spiral wound design. Then, preheated feed is heated to a required temperature before entering from its center. The feed solution flows axially in the evaporator channel, and the permeate which passes through the membrane flows spirally. The condenser channel is placed toward the shell side while feed towards the core side to get the advantage of counter-current circulation [6].

4. Performance evaluation

The MD performance metrics can be divided into two thermodynamic categories: local and system level. Local metrics are impacted by local properties such as porosity, pore size, thickness, membrane conductivity etc. These include permeate flux and thermal efficiency. On the other hand, the system-level metrics are impacted by the process parameters such as temperature, energy flow, etc. These can be divided into first law efficiencies (GOR, SEC) and second law efficiencies (thermal efficiency-II) [43]. In the context of MD, these can be described as follows:

1. *Permeate flux:* The permeate flux J (kg·m^{-2}h^{-2}) is the amount of distillate transported through a unit membrane area. It is the most significant parameter to evaluate the performance of an MD system. It can be expressed as:

$$J = \frac{\dot{m}_d}{A_m} \qquad (27)$$

Where A_m is the active surface area (m^2) of the membrane, and $\dot{m}_d$ is the amount of distillate that passes through the membrane (kg·s^{-1}).

2. *Gained output ratio:* Gained output ratio (GOR) is the first law efficiency of a thermal desalination system and is often used to quantify energy efficiency. It is defined as the ratio of thermal energy required to vaporize the distillate mass to actual heat input. Mathematically, it can be expressed as [44, 45]:

$$GOR = \frac{\dot{m}_d * h_{fg}}{Q_{in}} \qquad (28)$$

Where

$$Q_{in} = \dot{m}_f * c_p * \Delta T_{12} \qquad (29)$$

Where c_p is the specific heat capacity of fluid (Jkg^{-1}K^{-1}). For a system, with no heat energy recovery, GOR is simply a thermal efficiency without a heat exchanger. The GOR is a dimension-less quantity and its value lies between 0 and 1 for a single pass system without any heat recovery, and more than 1 if the evaporation and condensation heat is reused. In other words, GOR tells how many times the enthalpy of evaporation is reused. For most of the commercial systems and large distillation units, the condensation heat is utilized, therefore, shows values greater than 1.

3. *Specific energy consumption:* Specific energy consumption (SEC) is the energy consumed to produce a unit amount of distillate volume [42, 44, 46]. It is expressed as:

$$SEC = \frac{Q_{in}}{m_d} = \frac{\dot{m}_f * c_p * \Delta T_{12}}{m_d} \tag{30}$$

Where $\dot{m}_f$, c_p, ΔT_{12}, Q_{in}, and m_d are the mass flow rate of feed solution $(kg \cdot s^{-1})$, specific heat of water $(kJ \cdot kg^{-1}K^{-1})$, the temperature difference of inlet and outlet feed streams (K), the total input energy (kWh) consumed by the circulating feed, and the total mass of distillate produced (m^3), respectively.

4. *Thermal efficiency:* Thermal efficiency η_{th} in MD is the ratio of energy of distillate to that of actual energy used [44]. It is also called first law efficiency. It can be expressed as:

$$\eta_{th} = \frac{Q_v}{Q_v + Q_c} \tag{31}$$

$$\eta_{th} = \frac{J * A_m * h_{fg}}{Q_{in}} \tag{32}$$

Where J, A_m, h_{fg}, Q_m, Q_v, and Q_c are distillate flux $(kg \cdot m^{-2}h^{-1})$, membrane active area (m), latent heat of vaporization $(kJ \cdot kg^{-1}K^{-1})$, heat flux through membrane $(W \cdot m^{-2})$, heat energy of vaporization $(kJ \cdot kg^{-1}K^{-1})$ and heat of condensation $(kJ \cdot kg^{-1}K^{-1})$, respectively. The Q_{in} can be expressed as shown in Eq. (29)

To improve the thermal efficiency, the heat conduction Q_c should be minimized by using higher thickness membranes, an air gap etc. However, this may require optimizing other parameters as well. Thermal efficiency matches with the GOR if there is no heat recovery [43].

5. *Second law efficiency:* First law efficiency generally used to compare systems with same energy source. However, if the energy source is different, the two systems can not be compared fairly. Using second law efficiency, exergies are compared instead. Therefore, systems can be compared regardless of their energy source. Exergy is the maximum available work extracted from the system when the system moves from its initial state to equilibrium. **Figure 12** shows a representation of second law efficiency. The first law expression can be deduced as [47]:

$$\dot{Q}_H + (\dot{m} \ . h)_{sw} = \dot{Q}_0 + (\dot{m}.h)_d + (\dot{m}.h)_{br} \tag{33}$$

$$\dot{Q}_H - \dot{Q}_0 = (\dot{m}.h)_d + (\dot{m}.h)_{br} - (\dot{m} . h)_{sw} \tag{34}$$

The second law can be expressed as:

$$\frac{\dot{Q}_H}{T_0} - \frac{\dot{Q}_0}{T_0} = (\dot{m}.s)_d + (\dot{m}.s)_{br} - (\dot{m}.s)_{sw} + \dot{s}_{gen} \tag{35}$$

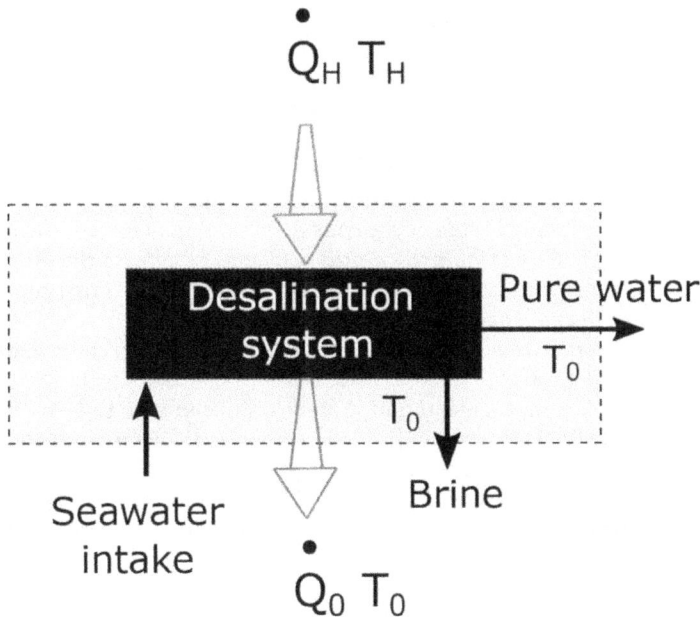

Figure 12.
A schematic diagram of a black-box desalination system with heat transfer occurring from an external heat source at temperature T_H, and to the environment at temperature T_o.

Multiplying second law equation Eq. (35) by T_0 and subtracting from Eq. (34):

$$\left(1 - \frac{T_0}{T_H}\right)\frac{\dot{Q}_H}{\dot{m}_d} = (G_d - G_{br}) - \frac{1}{RR}(G_{sw} - G_{br}) + T_0\frac{\dot{S}_{gen}}{\dot{m}_d} = \frac{\dot{W}_{sep}}{\dot{m}_d} \qquad (36)$$

Where RR is the recovery ratio $\frac{\dot{m}_p}{\dot{m}_{sw}}$, G is the Gibbs free energy $(h - T.s)$. The second law efficiency can be expressed as:

$$\eta_{II} = \frac{\dot{W}_{separation}}{\dot{W}_{used}} \qquad (37)$$

6. *Recovery ratio:* Recovery rate (RR) is the distillate production relative to the input feed stream flow in the MD system [48]. It is expressed as:

$$RR = \frac{\dot{m}_d}{\dot{m}_f} * 100\% \qquad (38)$$

A high recovery means a high distillate flow rate is obtained by a given feed flow. In a single pass MD system, feed recovery is very low as compared to other membrane systems. It is reported that the maximum recovery attained in single pass MD reached 10% even if 100% thermal efficiency is attained [49].

In the latest developments, hybridization of MD with existing technologies is used to improve the energy efficiency of MD. The MD has been successfully integrated with other processes as hybrid such as RO, MED, and FO. MD unit has been successfully realized to treat RO brine and FO draw solution which is a challenging part of the process. A comprehensive review of different hybrid technologies with the MD is described by Ghaffour et al. [50]. Additionally, in-situ heating of the feed water inside the MD module, so-called localized heating, has been introduced recently. Localized heating has shown a decreased TP effect which ensures the delivery of heat energy at the site of the feed-membrane interface. It eliminates the circulation heat loss associated with the conventional bulk heating [17, 51]. Photothermal energy source is the prime consideration due to its renewable and ever-existing energy source where the heat is delivered directly to the MD membrane. Politano et al. introduced the photothermal concept in MD first time using surface plasmon effect of silver nanoparticles [52, 53]. Various organic, inorganic, and polymeric materials have also been investigated as photothermal materials in the MD system [6, 54–56].

Similarly, Joule heating elements and spacers have been used to deliver localized heating to the feed channel [57, 58]. Ahmed et al [17] recently demonstrated the TP reduction by using the electrothermal property of carbon nanostructure. They obtained a decrease of SEC by 58%. Hence, localized heating provides a relatively simple infrastructure for small-scale clean water generation in remote off-grid regions.

5. Conclusions

Membrane distillation (MD) is a promising technology for the separation and purification industry. It is a specific distillation process in which vapor molecules travel through a hydrophobic membrane. MD has several advantages, including low-grade heat input, less fouling propensity, ability to treat high saline water. Four typical membrane configurations, membrane characteristics, membrane modules, heat and mass transfer mechanisms, thermal efficiency, and operating parameters have been presented. The most important limitation that has to be considered with membrane distillation is temperature polarization, which reduces the trans-membrane temperature difference, hence the performance. MD is found to be most suitable when the input energy source is solar or waste heat, due to energy intensive nature of distillation process. The recent development in membrane technology allows MD to run in compact modular configurations such as spiral module configurations. The performance of MD is expressed in terms of flux output and the specific energy. Although flux is an important but not the only factor to demonstrate the performance of an MD system, energy input also plays an important role. The recent advances in localized heating make the MD more promising to operate on a bigger scale.

Acknowledgements

The authors would like to acknowledge King Abdullah University of Science and Technology (KAUST) for supporting the authors in preparing this manuscript.

Conflict of interest

The authors declare no competing financial interests.

Nomenclature

AGMD	Air gap membrane distillation
CAPEX	Capital expenditure
DCMD	Direct contact membrane disitillation
EE	Energy efficiency
FS	Flat sheet
GOR	Gain output ratio
MD	Membrane distillation
OPEX	Operational expenditure
RO	Reverse osmosis
RR	Recovery ratio
SEC	Specific energy consumption
SEM	Scanning electron microscope
SGMD	Sweeping gas membrane distillation
TP	Temperature polarization
TDS	Total dissolved solids
TPC	Temperature polarization coefficient
VMD	Vacuum membrane distillation

Symbols

A_m	membrane surface area (m^2)
C	concentration ($mol \cdot L^{-1}$)
C_m	membrane mass transfer coefficient ($kg \cdot m^{-2} \cdot s^{-1} \cdot Pa^{-1}$)
c_P	specific heat capacity ($J \cdot kg^{-1} \cdot K^{-1}$)
α	water activity ($-$)
χ	membrane tortuosity ($-$)
δ	membrane thickness (M)
$\dot{m}$	mass flow rate ($kg \cdot s^{-1}$)
$\dot{Q}$	heat transfer rate (W)
η	efficiency ($-$)
γ	activity coefficient ($-$)
λ	molecular mean free path (m)
μ	viscocity ($Pa \cdot s$)
π	osmotic pressure (Pa)
Π	pi (3.14) ($-$)
ρ	density ($kg \cdot m^{-3}$)
σ	molecular collision diameter (m)
Θ	temperature polarization ($-$)
ε	membrane porosity ($-$)
D_p	diameter of pore (m)
D_w	diffusion coefficient of water ($m^2 \cdot s^{-1}$)
G	Gibbs free energy ($kJ \cdot kg^{-1}$)
h	heat transfer coefficient ($W \cdot m^{-2} K^{-1}$)
h_{fg}	latent heat of vaporization ($kJ \cdot kg^{-1}$)
J	vapor flux ($kg \cdot m^{-2} \cdot h^{-1}$)
k_B	Boltzmann constant ($J \cdot K^{-1}$)
k_m	thermal conductivity ($W \cdot m^{-1} K^{-1}$)
K_n	Knudsen number ($-$)
M	molecular weight ($kg \cdot mol^{-1}$)

$\dot{m}$	mass flow rate (mg·s^{-1})
N_A	Avagadro's number (6.023×10^{23}) (mol^{-1})
Nu	Nusselt number $(-)$
p	partial pressure (Pa)
P	total pressure (Pa)
Pr	Prandtl number $(-)$
q	heat flux (W·m^2)
$\dot{Q}$	heat transfer rate (kW)
r	pore radius (m)
R	universal gas constant (J·mol^{-1}K^{-1})
Sc	Schmidt number $(-)$
s	specific entropy generation (kJ·kg^{-1}K^{-1})
T	temperature (K)
t	time (s)
U	overall heat transfer coefficient (W·m^{-2}·K^{-1})
v	specific volume (m^3·kg^{-1})
W	work (kW)
X	mole fraction $(-)$

Subscripts

0	natural state $(-)$
12	between input and output $(-)$
av	average $(-)$
b	bulk $(-)$
br	brine $(-)$
c	conduction $(-)$
d	distillate $(-)$
f	feed $(-)$
fb	bulk feed $(-)$
fg	fluid-gas $(-)$
g	gaseous state $(-)$
gen	generation $(-)$
H	heat $(-)$
m	membrane $(-)$
p	permeate $(-)$
pb	bulk permeate $(-)$
sw	seawater $(-)$
sat	saturation $(-)$
v	vapor $(-)$

Desalination by Membrane Distillation
DOI: http://dx.doi.org/10.5772/intechopen.101457

Author details

Mustakeem Mustakeem*, Sofiane Soukane, Muhammad Saqib Nawaz
and Noreddine Ghaffour
Biological and Environmental Science and Engineering (BESE) Division, Water
Desalination and Reuse Center (WDRC), King Abdullah University of Science and
Technology (KAUST), Thuwal, Saudi Arabia

*Address all correspondence to: mustakeem.mustakeem@kaust.edu.sa

IntechOpen

References

[1] Lawson KW, Lloyd DR. Membrane distillation. Journal of Membrane Science. 1997;**124**(1):1-25

[2] Johnson RA, Nguyen MH. Understanding Membrane Distillation and Osmotic Distillation. New York: John Wiley & Sons; 2017

[3] Burgoyne A, Vahdati M. Direct contact membrane distillation. Separation Science and Technology. 2000;**35**(8):1257-1284

[4] Belessiotis V, Kalogirou S, Delyannis E. Chapter four–membrane distillation. Thermal Solar Desalination: Methods and Systems. 2016;**201**:191-251

[5] Alkhudhiri A, Hilal N. Membrane distillation—Principles, applications, configurations, design, and implementation. In: Emerging Technologies for Sustainable Desalination Handbook. Amsterdam, Netherlands: Elsevier; 2018. pp. 55-106

[6] Zhani K, Zarzoum K, Ben Bacha H, Koschikowski J, Pfeifle D. Autonomous solar powered membrane distillation systems: State of the art. Desalination and Water Treatment. 2016;**57**(48–49): 23038-23051

[7] Alkhudhiri A, Darwish N, Hilal N. Membrane distillation: A comprehensive review. Desalination. 2012;**287**:2-18

[8] Kebria MRS, Rahimpour A. Membrane Distillation: Basics, Advances, and Applications. Rijeka: IntechOpen; 2020

[9] Narayan A, Pitchumani R. Analysis of an air-cooled air gap membrane distillation module. Desalination. 2020; **475**:114179

[10] Al-Zoubi H, Al-Amri F, Khalifa AE, Al-Zoubi A, Muhammad A, Ebtehal Y, et al. A comprehensive review of air gap membrane distillation process. Desalination and Water Treatment. 2018;**110**:27-64

[11] Khayet M, Matsuura T. Chapter 11— Sweeping Gas Membrane Distillation. Membrane Distillation: Principles and Applications. Amsterdam, The Netherlands: Elsevier; 2011

[12] Suárez F, Tyler SW, Childress AE. A theoretical study of a direct contact membrane distillation system coupled to a salt-gradient solar pond for terminal lakes reclamation. Water Research. 2010;**44**(15):4601-4615

[13] Gryta M, Tomaszewska M, Morawski A. Membrane distillation with laminar flow. Separation and Purification Technology. 1997;**11**(2): 93-101

[14] Drioli E, Ali A, Macedonio F. Membrane distillation: Recent developments and perspectives. Desalination. 2015;**356**:56-84

[15] Khayet M, Matsuura T. Chapter 10—Direct Contact Membrane Distillation. Membrane Distillation: Principles and Applications. Amsterdam, The Netherlands: Elsevier; 2011

[16] Thomas N, Mavukkandy MO, Loutatidou S, Arafat HA. Membrane distillation research & implementation: Lessons from the past five decades. Separation and Purification Technology. 2017;**189**:108-127

[17] Ahmed FE, Lalia BS, Hashaikeh R, Hilal N. Alternative heating techniques in membrane distillation: A review. Desalination. 2020;**496**:1-14

[18] Al-Obaidani S, Curcio E, Macedonio F, Di Profio G, Al-Hinai H, Drioli E. Potential of membrane

distillation in seawater desalination: thermal efficiency, sensitivity study and cost estimation. Journal of Membrane Science. 2008;**323**(1):85-98

[19] Olatunji SO, Camacho LM. Heat and mass transport in modeling membrane distillation configurations: A review. Frontiers in Energy Research. 2018;**6**:130

[20] Summers EK, Arafat HA, Lienhard JH. Energy efficiency comparison of single-stage membrane distillation (md) desalination cycles in different configurations. Desalination. 2012;**290**:54-66

[21] Anvari A. Mitigation of Temperature Polarization and Mineral Scaling in Membrane Distillation: The Impact of Induction Heated Elements. PhD thesis, Temple University; 2020

[22] Camacho LM, Dumée L, Zhang J, Li J-d, Duke M, Gomez J, et al. Advances in membrane distillation for water desalination and purification applications. Water. 2013;**5**(1):94-196

[23] Schofield R, Fane A, Fell C. Heat and mass transfer in membrane distillation. Journal of Membrane Science. 1987;**33**(3):299-313

[24] Rizvi SS, Pabby AK, Sastre AM. Handbook of Membrane Separations: Chemical, Pharmaceutical, Food, and Biotechnological Applications. London: CRC Press; 2009

[25] L. Martinez-Diez, M. Vázquez-González, and F. Florido-Diaz, Temperature Polarization Coefficients in Membrane Distillation. Taylor & Francis; 1998

[26] Alsaadi AS, Francis L, Amy GL, Ghaffour N. Experimental and theoretical analyses of temperature polarization effect in vacuum

membrane distillation. Journal of Membrane Science. 2014;**471**: 138-148

[27] Martinez D, Vazquez-Gonzalez M, Florido-Diaz F. Study of membrane distillation using channel spacers. Journal of Membrane Science. 1998;**144**: 45-56

[28] Velazquez A, Mengual JI. Temperature polarization coefficients in membrane distillation. Industrial & engineering chemistry research. 1995; **34**(2):585-590

[29] Kim Y-D, Thu K, Ghaffour N, Ng KC. Performance investigation of a solar-assisted direct contact membrane distillation system. Journal of Membrane Science. 2013;**427**:345-364

[30] Curcio E, Drioli E. Membrane distillation and related operations—A review. Separation & Purification Reviews. 2005;**34**(1):35-86

[31] Kim Y-D, Francis L, Lee J-G, Ham M-G, Ghaffour N. Effect of non-woven net spacer on a direct contact membrane distillation performance: Experimental and theoretical studies. Journal of Membrane Science. 2018;**564**:193-203

[32] Seo J, Kim YM, Kim JH. Spacer optimization strategy for direct contact membrane distillation: Shapes, configurations, diameters, and numbers of spacer filaments. Desalination. 2017; **417**:9-18

[33] Shakaib M, Hasani SMF, Ahmed I, Yunus RM. A CFD study on the effect of spacer orientation on temperature polarization in membrane distillation modules. Desalination. 2012;**284**: 332-340

[34] Taamneh Y, Bataineh K. Improving the performance of direct contact membrane distillation utilizing spacer-filled channel. Desalination. 2017;**408**: 25-35

[35] Chen X, Vanangamudi A, Wang J, Jegatheesan J, Mishra V, Sharma R, et al. Direct contact membrane distillation for effective concentration of perfluoroalkyl substances: Impact of surface fouling and material stability. Water Research. 2020;**182**:116010

[36] Alklaibi AM, Lior N. Membrane-distillation desalination: Status and potential. Desalination. 2005;**171**(2): 111-131

[37] Martinez L, Rodriguez-Maroto JM. On transport resistances in direct contact membrane distillation. Journal of Membrane Science. 2007;**295**(1–2): 28-39

[38] Martinez-Diez L, Florido-Diaz F. Desalination of brines by membrane distillation. Desalination. 2001;**137**(1–3): 267-273

[39] Khayet M, Matsuura T. Chapter 10—Direct Contact Membrane Distillation. Amsterdam, Netherlands: Elsevier; 2011. pp. 249-293

[40] Lee J-G, Alsaadi AS, Karam AM, Francis L, Soukane S, Ghaffour N. Total water production capacity inversion phenomenon in multi-stage direct contact membrane distillation: A theoretical study. Journal of Membrane Science. 2017;**544**:126-134

[41] Ali A, Criscuoli A, Macedonio F, Drioli E. A comparative analysis of flat sheet and capillary membranes for membrane distillation applications. Desalination. 2019;**456**:1-12

[42] Winter D, Koschikowski J, Wieghaus M. Desalination using membrane distillation: Experimental studies on full scale spiral wound modules. Journal of Membrane Science. 2011;**375**(1–2):104-112

[43] Warsinger DM, Nejati S, Juybari HF. Energy efficiency metrics in membrane distillation. In: Advances in Water Desalination Technologies. Singapore: World Scientific; 2021. pp. 263-288

[44] Khayet M. Solar desalination by membrane distillation: Dispersion in energy consumption analysis and water production costs (a review). Desalination. 2013;**308**:89-101

[45] Qtaishat MR, Banat F. Desalination by solar powered membrane distillation systems. Desalination. 2013;**308**:186-197

[46] Jantaporn W, Ali A, Aimar P. Specific energy requirement of direct contact membrane distillation. Chemical Engineering Research and Design. 2017; **128**:15-26

[47] Warsinger DEM. Thermodynamic Design and Fouling of Membrane Distillation Systems. PhD thesis, Massachusetts Institute of Technology; 2015

[48] Saffarini RB, Summers EK, Arafat HA, et al. Technical evaluation of stand-alone solar powered membrane distillation systems. Desalination. 2012; **286**:332-341

[49] Lokare OR, Tavakkoli S, Khanna V, Vidic RD. Importance of feed recirculation for the overall energy consumption in membrane distillation systems. Desalination. 2018;**428**:250-254

[50] Ghaffour N, Soukane S, Lee J-G, Kim Y, Alpatova A. Membrane distillation hybrids for water production and energy efficiency enhancement: A critical review. Applied Energy. 2019; **254**:113698

[51] Wu X, Jiang Q, Ghim D, Singamaneni S, Jun Y-S. Localized heating with a photothermal polydopamine coating facilitates a novel membrane distillation process. Journal of Materials Chemistry A. 2018;**6**(39): 18799-18807

[52] Politano A, Di Profio G,
Fontananova E, Sanna V, Cupolillo A,
Curcio E. Overcoming temperature
polarization in membrane distillation by
thermoplasmonic effects activated by ag
nanofillers in polymeric membranes.
Desalination. 2019;**451**:192-199

[53] Politano A, Argurio P, Di Profio G,
Sanna V, Cupolillo A, Chakraborty S,
et al. Photothermal membrane
distillation for seawater desalination.
Advance Material. 2017;**29**(2):1-6

[54] Razaqpur AG, Wang Y, Liao X,
Liao Y, Wang R. Progress of
photothermal membrane distillation for
decentralized desalination: A review.
Water Research. 2021;**201**:117299

[55] Huang L, Pei J, Jiang H, Hu X. Water
desalination under one sun using
graphene-based material modified PTFE
membrane. Desalination. 2018;**442**:1-7

[56] Ghim D, Wu X, Suazo M, Jun Y-S.
Achieving maximum recovery of latent
heat in photothermally driven multilayer stacked membrane distillation.
Nano Energy. 2021;**80**:1-10

[57] Mustakeem M, Qamar A,
Alpatova A, Ghaffour N. Dead-end
membrane distillation with localized
interfacial heating for sustainable and
energy-efficient desalination. Water
Research. 2021;**189**:116584

[58] Tan YZ, Ang EH, Chew JW.
Metallic spacers to enhance membrane
distillation. Journal of Membrane
Science. 2019;**572**:171-183

Chapter 5

Modeling of Solar-Powered Desalination

Zafar Abbas, Nasir Hayat, Anwar Khan
and Muhammad Irfan

Abstract

The scarcity, global, and local demand of pure water for SDGs become promi-
nent issue. The global emissions of CO2 and GHGs have put pressure to develop the
solar-powered desalination plants. This article discussed the selection of site for the
solar thermal desalination in Pakistan keeping the eye on sustainability and modeling and cost analysis of single solar stills technology at Lyari River in Karachi,
Pakistan. Pakistan is among the water-deficit countries having 35% of population
having lack of pure drinkable water. The plenty of solar irradiance and saline water
in Pakistan make it very favorable for solar-powered desalination. The solar stills
technology is one of the best technologies to meet the local demand of pure water.
The modeling is composed of governing equations based on the law of conservation
of mass and law of conservation of energy. The solar irradiance at Lyari River is
taken from MERRA–2. The result depicted that the hourly production of distill
water is $1\,kg/m^3$ and $8\,kg/m^3$ with and without the FRL lens. The cost of distill water
produced from the solar stills having FRL lens is 33% less as compared with solar
stills without FRL lens.

Keywords: SDGs, GHGs, desalination, solar powered, solar stills

1. Introduction

Water is the resource that sustains all life on the planet earth and key element of
sustainable development [1]. The rapid growth in population, and industrial and
economic development needs high demand of water. The need of freshwater for
drinking and potable water in arid areas is increasingly important issues in most
part of the world. In 2000, the world annual demand for water is 4000 billion cubic
meter. By 2030, it is estimated to increase over 58% [2]. Water availability per
person in Pakistan was 5,600 cubic meter in 1960, and it is reduced to 1000 cubic
meter in 2018. The demand of water in Pakistan is important because of its agrarian
nature of economy and the agriculture sector shares 24% of gross domestic product
(GDP). The regional conflicts on the availability and use of water have pressure on
the demand of water. The water sources in Pakistan are surface water, rainfall,
glaciers, and groundwater. Surface water consists of rivers, lakes, dams, and runoff
during and after heavy rains. Mostly, the groundwater is the source in urban areas
except in Karachi, Hyderabad, and some part of Islamabad use surface water. Water
for rural areas is also from groundwater source except in saline groundwater areas
where irrigation canals are used for domestic purpose [3]. Currently, the water

availability per capita in Pakistan is 1000 cubic meter. According to Population Action International, 1993, the countries with water availability below 1000 cubic meter experience chronic water stress [1]. Presently, more than 65% people of total population have access to safe drinking water including 85 and 55% urban and rural areas, respectively. The 35% of population has lack of drinkable water in Pakistan [3]. According to WHO, a drinkable water should have dissolved salt concentration less than 500 ppm. The normal seawater and brackish water have dissolved salt and ion concentration of 3500 ppm and 1000 ppm. Therefore, desalination of seawater and brackish water is the way to make the water drinkable. Most of the desalination plants use conventional methods of energy. But the fossil fuel methods of energy sources have adverse impact on environmental sustainability by producing air pollution, global warming, and GHGs emission. The utilization of fossil fuels for the desalination plants is contributing in CO_2 emissions. The total installed power plant for the desalination processes is responsible for the emission of 76 million tons (Mt) of carbon dioxide per year. In 2040, the emission of CO_2 is expected to 218 million tons per year [4]. The cuse of fossil fuels for desalination plant is neither sustainable nor environment friendly. Therefore, there is a need of alternative sources of energy to achieve the world demand of freshwater. At a same time, the alternative source should be sustainable and environmental friendly. The renewable energy sources of energy are the alternatives to power desalination processes. Thus, the solar power desalination is one of the most suitable alternatives for desalination plant that meets water demand and also environmental friendly.

Therefore, in this research paper the focus is on the demand of water in Pakistan as the result of rapid growth in population and industrialization. It has become necessary to install the desalination plant in Pakistan by keeping in mind the energy available as well as economic situation. The main ambitions of this research are to select a site having plenty of solar radiations and salt or brackish and suitable solar technology having low capital and operational cost to fulfill the demand of pure water at minimum cost. Thus, the development of mathematical model of solar stills and cost analysis at Lyari River, in Karachi, follows the solution of mathematical model using MATLAB.

2. Methodology

The methodology of this research is composed of the selection of site in Pakistan for solar power desalination following the mathematical modeling of the single slope of solar stills and employs modern software for the solution of mathematical model. The governing equations for mathematical model of the solar stills are based on the law of conservation of mass and law of conservation of energy for the system. The equations for convective heat transfer coefficients and radiation heat transfer coefficients are based on the Dunkle's model. The MATLAB r2019 is employed to solve the equations.

2.1 Site selection in Pakistan

The availability of **1900–2200 kWh/m²** annual global irradiance makes Pakistan highly favorable for solar power-based desalination [5]. The Balochistan and Sindh province of Pakistan is rich in solar energy with an average daily direct normal irradiance of 5.3–5.6 kWh/m² and 2.5–3.0 kWh/m² with sunshine duration of 8–8.5 hours a day [6]. Therefore, in this research paper the site at Lyari River Karachi has been selected for the modeling the single-slope solar stills.

2.2 Mathematical modeling of solar stills

The basic assumptions while modeling the solar stills take negligible temperature stratification within the evaporator basin. Temperature is uniform within each still component. Temperature is time dependent. The evaporated water is assumed only pure water; that is, the evaporated water has no dissolved salt or ion. The stills have no vapor leakages. The governing equations are based on law of conservation of mass and law of conservation of energy. The schematic of single slope solar stills is shown in **Figure 1**.

The law of conservation of mass can be written as [7].

$$\dot{m}_{sw} = \dot{m}_{ev} + \dot{m}_b \tag{1}$$

If **Xsw** is the concentration of salt in the feed saline water, **Xb** is the concentration of salt in the cbrine within the basin. Then, the salt balance is [7].

$$\dot{m}_{sw}X_{sw} = \dot{m}_bX_b \tag{2}$$

The solubility of salt determines the salt content in the brine. The salt content in the brine is important in practice to avoid the problem of forming layer and blockage. The factor fc for concentration is defined as the ratio of brine concentration to feed concentration.

$$fc = \frac{X_b}{X_{sw}} \tag{3}$$

This factor is used to fix a threshold limit to not exceed during evaporation and condensation. By solving Eq. (2)) and Eq. (3), we have the following equation.

$$\dot{m}_b = \frac{1}{fc}\dot{m}_{sw} \tag{4}$$

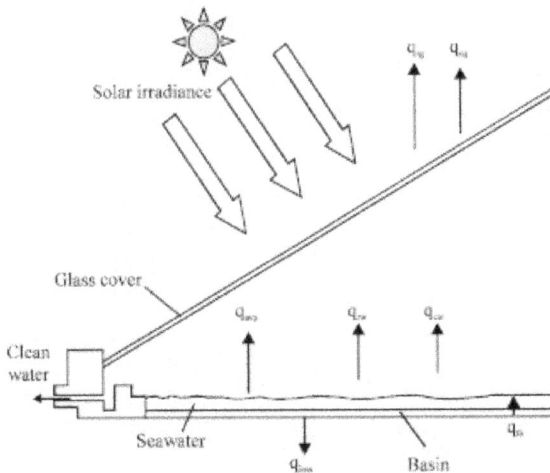

Figure 1.
Schematic of single-slope solar stills.

$$\dot{m}_{ev} = \frac{fc - 1}{fc} \dot{m}_{sw} \tag{5}$$

Eq. (5) is for the stationary conditions, the rate of evaporated water as a function of rate of feed saline water. Now, the distillated water or recovery rate can be defined as

$$\varphi = \frac{fc - 1}{fc} \tag{6}$$

The recovery rate is an important parameter, which indicates the possible amount of distillate water from the saline feed water without scaling [7]. It means that only 40% of saline water can be transformed into distillate water without encrustation and blockage.

The law of conservation of energy gives the following set of equations for the respective components in the solar stills.

The energy balance equation for the outer of the transparent glass cover is as follows [7]:

$$\frac{\rho_g V_g C_{pg}}{2A_g} \frac{dT_{ge}}{dt} = \frac{\lambda_g}{e_g} \left(T_{gi} - T_{ge} \right) - hc_{ge-am} \left(T_{ge} - T_{amb} \right) - hr_{ge-sky} \left(T_{ge} - T_{sky} \right) \tag{7}$$

The energy balance equation for the inner of the transparent glass cover is as follows:

$$\frac{\rho_g V_g C_{pg}}{2A_g} \frac{dT_{ge}}{dt} = \acute{a}_g I(t) + \left(hc_{sw-gi} + hr_{sw-gi} \right) \left(T_{sw} - T_{gi} \right) + \frac{\dot{m}_{ev} L_v}{A_g} - \frac{\lambda_g}{e_g} \left(T_{gi} - T_{ge} \right) \tag{8}$$

The energy balance for the seawater inside the basin of solar still is as follows [7]:

$$\frac{\rho_{sw} V_{sw} C_{p,sw}}{A_{sw}} \frac{dT_{sw}}{dt} =$$

$$\acute{a}_{sw} I(t) + hc_{c-sw} \left(T_c - T_{sw} \right) + \frac{\rho_{sw} D_{sw}}{A_{sw}} \left(c_{p,sw} T_{sw,in} - c_{p,b} T_{b,out} \right) - \frac{\dot{m}_{ev}}{A_{sw}} \left(h_L - c_{p,b} T_{b,out} \right)$$
$$- \left(hc_{sw-ig} - hr_{sw-ig} \right) \left(T_{sw} - T_{gi} \right) \tag{9}$$

The convective heat transfer coefficient of between the outer of the transparent glass cover and the ambient temperature depends on the wind velocity. According to McAdams correlation [8], this coefficient is approximated by the following equation.

$$hc_{ge-amb} = \begin{cases} 5.621 + \dfrac{1151.2 v}{T_{amb}} \: if \: v < 4.88 \: ms^{-1} \\ 604.29 \left(\dfrac{v}{T_{amb}} \right)^{0.78} \: if \: 4.88 \leq v < 30.48 \: ms^{-1} \end{cases} \tag{10}$$

The heat transfer coefficient between the saline water and the inner of the transparent glass cover is given by the second form of Dunkle's model and can be written as [9].

$$hc_{sw-gi} = 0.884 \left(T_{sw} - T_{gi} + \frac{\left(p_{sw} - p_{gi}\right) T_{sw}}{2.689 \times 10^5 - p_{sw}} \right)^{\frac{1}{3}} \tag{11}$$

The radiation heat transfer coefficient between the outer of the transparent glass cover and the sky is given by

$$hr_{ge-sky} = \varepsilon_v \sigma \left(T^2_{ge} + T^2_{sky}\right)\left(T_{ge} + T_{sky}\right) \tag{12}$$

σ is the Steffen Boltzman constant.
The sky temperature is determined by [10].

$$T_{sky} = T_{amb}(0.74 + 0.006\theta)^{0.25} \tag{13}$$

Where Θ is the dew point temperature given by [11].

$$\theta = \frac{273.3}{17.27 - In\varepsilon + \frac{17.27 T_{amb} - 4061}{T_{amb} - 35.85}} \left(In\varepsilon + \frac{17.27 T_{amb} - 4061}{T_{amb} - 35.85} \right) \tag{14}$$

ε is the relative humidity.
The radiation heat transfer coefficient between the saline water and the sky is expressed as

$$hr_{ge-sky} = \varepsilon_{eff} \sigma \left(T^2_{sw} + T^2_{gi}\right)\left(T_{sw} + T_{gi}\right) \tag{15}$$

Emissivity is given by,

$$\varepsilon_{eff} = \left(\frac{1}{\varepsilon_{sw}} + \frac{1}{\varepsilon_g} - 1 \right)^{-1} \tag{16}$$

The equation in the second part of the (Eq. (9)) in given by Dunkle's model as [9].

$$\dot{m}_{ev} h_L = h_{ev} A_{sw} \left(T_{sw} - T_{gi}\right) \tag{17}$$

The latent heat of vaporization h_L is given by [9].

$$h_L = 3146 - 2.36 T_{sw} \tag{18}$$

The evaporative heat transfer coefficient h_{ev} is given by [9].

$$h_{ev} = 0.016273 \, hcv_{sw-gi} \frac{p_{sw} - p_{gi}}{T_{sw} - T_{gi}} \tag{19}$$

The evaporative heat transfer coefficients Eq. (11) and Eq. (19) can only be estimated through correlations when the following conditions is satisfied: the aspect ratio $2.5 \leq a \leq 5.5$, the inclination angle $10^0 \leq i \leq 30^0$, and Rayleigh number $5 \times 10^6 \leq Ra \leq 5 \times 10^7$.

If the above conditions are not fulfilled, then it could be done either experimentally or by using 2D modeling of the problem such as considered in some other systems [12, 13].

The Nusselt number is obtained through a correlation in the form by to express the convective heat transfer coefficients [14].

$$Nu = c(Ra)^n = c(GrPr)^n \tag{20}$$

The Grashof and Prandtl number is given by,

$$Gr = \frac{\beta g \rho^2 L^3 \triangle T}{\mu^2} \tag{21}$$

$$Pr = \frac{\mu c_\rho}{\lambda} \tag{22}$$

The correlation that gives Nusselt number is [14].

$$Nu = \begin{cases} 1 \ if Gr < 10^5 \\ 0.5(Ra)^{0.25} if \ 10^5 < Gr < 2 \times 10^7 \\ 0.15(Ra)^{0.33} if Gr > 2 \times 10^7 \end{cases} \tag{23}$$

When the Nusselt number is known, the heat transfer coefficient between the basin liner plate and the saline water can be calculated for an active solar still as

$$hcv_{c-sw} = \frac{Nu_{c-sw}\lambda_{sw}}{L} \tag{24}$$

The convective heat transfer coefficient between the fluid and the plate for an active solar still is calculated as [15].

$$hcv_{h-c} = \frac{Nu_{h-c}\lambda_h}{L} \tag{25}$$

The heat loss coefficient is approximated by the following equation [15].

$$U_{loss} = \frac{\lambda_{is}}{e_{is}} \tag{26}$$

$T_{b,out}$ can be assumed to be equal to that of the plate T_p for larger length of basin liner for an active solar stills [15].

The MATLAB's solver for ordinary differential equations (ODEs), MATLAB ode45 function, has been employed for the efficient computation of the differential equations.

The material properties and dimensions of solar still are given in **Table 1** and thermophysical properties of glass, basin, and insulation are given in **Table 2** in Appendix A.

3. Results

The hourly production of distill water in *kg/hour* at Lyari River, Karachi, with and without Fresnel (FRL) lens. The results are depicted in **Figure 2**.

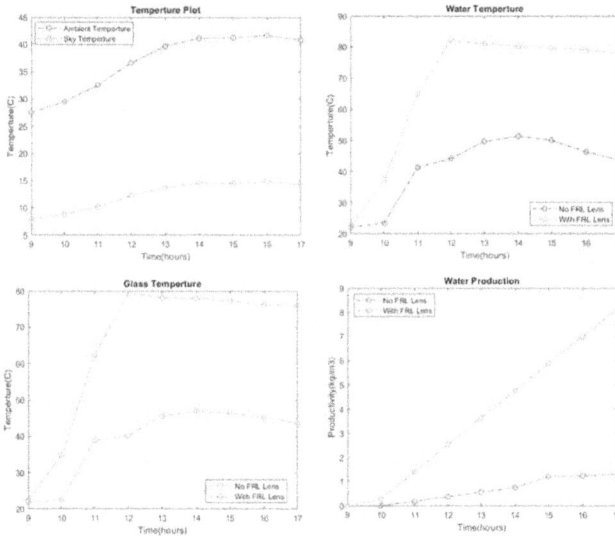

Figure 2.
Result of mathematical model of single slope solar still.

The maximum ambient temperature and sky temperature on the hottest day is 39.5°C and 14.7°C. The result is showing that the maximum water temperature with and without Fresnel (FRL) lens is 82.3°C and 47.2°C. And also the maximum glass temperature with and without FRL lens is found to be 80°C and 39.5°C. The production of water is calculated using the temperatures. The maximum water production with and without FRL lens is 8 kg/hour and 1 kg/hour. Using FRL lens, the production of water is 330% more than without using FRL lens.

4. Cost analysis

The cost analysis of the solar still with and without FRL lens includes the capital, operational, and maintenance cost. The major contributor is FRL lens that cost 90$. The details of the costs are given in Appendix A (**Table 3**).

The economic performance is estimated by the following.

$$P = \frac{Capital\ cost + Operational\ cost/year + Maintenance\ cost/year}{water\ production/year} \qquad (27)$$

The monthly operating cost is about 1.25$. There is no maintenance cost is required in this case but only the cleaning cost. The accidental cost is not considered in this study. The cost with FRL is 122.3$ and without FRL is 22.27$. The production of distilled water per cubic meter with and without FRL is found to be 1.37$ and 1.66$, respectively.

5. Conclusion

It is concluded that the scarcity of pure water can be compensated by desalination processes to meet the global demand of water as some developing countries have already done. The developed and developing countries have the capacity to install the conventional source of desalination plant but this attitude is greatly impacting on the environmental issues such as global emissions of CO_2 and greenhouse gases (GHGs). Currently, the desalination plants are based on the conventional sources of thermal energy. The sustainable development goals (SDGs) can only be achieved using the renewable source of energy for the desalination processes. This will eliminate two main problems: global emissions and scarcity of water. The alternative and most effective source for desalination processes keeping in mind the SDGs is the solar thermal desalination processes. The capital and operational cost of the conventional thermal desalination processes are high enough that under developing countries cannot afford it. Therefore, the solar thermal desalination processes are the best option for those countries. The plenty of solar irradiance, water, and land make Pakistan the best suited area for the solar thermal desalination. Baluchistan, Sindh, and Southern Punjab are the most suitable area for the solar energy applications. The Lyari River at Karachi in Sindh province is one of the most suitable areas for the solar thermal desalination processes. The solar stills technology for the distillation of saline water is one of the most favorable technologies to distill water to meet the water demand of Pakistan at effective cost. The governing equations of mathematical modeling of solar still were based on law of conservation of mass and law of conservation of energy. MATLAB was employed to solve the governing equations of the mathematical model. The result is showing that the utilization of Fresnel (FRL) lens makes the solar stills more productive of distill water as compared with solar stills without Fresnel lens. At the same time, the cost of pure water is less while using FRL lens in the solar stills. The solar stills technology works more efficiently at the remote areas of Pakistan where high-cost desalination plants are far enough to install. And ease of installation, capital, operational, and maintenance cost make it possible to reach to all people.

Acknowledgements

I pay gratitude to Allah Almighty for His blessings that make possible to complete this research paper. Our heartedly gratitude to our parents and guidance that they support us from childhood to present. Besides their support, it was impossible for us to do so.

It is my pleasure to acknowledge here the personal and institutional support I have received leading to the completion of this work. I am thankful to my team members, Muhammad Irfan and Anwar Khan, student of B.Sc. Mechanical Engineering Department at University of Engineering and Technology, Lahore, for their cooperation.

On the behalf of my team I would like to express sincere gratitude to Professor Nasir Hayat, Chairman Mechanical Engineering Department, University of Engineering and Technology Lahore for his continuous supervision, advice, effort, and worthy suggestions during the entire research project. At Mechanical Engineering Department, University of Engineering and Technology, Lahore, I am grateful to project coordinator Dr. Naseer Ahmad, Associate Professor Mechanical Engineering Department, and his fellows for their valuable and estimable suggestions. And special thanks go to our semester coordinator Dr. Zia ur Rahman Tahir.

I would like to thanks all the researchers and coordinators of the websites that their research articles are easily available at the respective websites. It is my pleasure that to express thank to Maja Bozecevic, Author Service Manager at IntechOpen, which is world-leading publisher of Open Book Access.

Conflict of interest

There is no conflict of interest for this publication.

Nomenclature

Symbols

A_{sw}	Area of saline water (m^2)
A_g	Area of glass cover (m^2)
A_b	Area basin (m^2)
A	Aspect ratio (dimensionless)
c	Constant in Nusselt correlation (dimensionless)
$c_{p,b}$	Specific heat of brine ($J\ kg^{-1}\ K^{-1}$)
$c_{p,sw}$	Specific heat of saline water ($J\ kg^{-1}\ K^{-1}$)
$c_{p,g}$	Specific heat of glass ($J\ kg^{-1}\ K^{-1}$)
e_g	Thickness of glass cover (mm)
e_{gs}	Thickness of insulation material (mm)
Fc	Feed concentration factor (dimensionless)
I(t)	Solar Intensity (Wm^{-2})
I_O	Constant solar intensity (Wm^{-2})
I	Inclination angle of glass cover (degree)
Gr	Grash of number
G	Acceleration of gravity ($m\ s^{-2}$)
Hl	Height of the higher side of the still (m)
Hr	Height of the lower side of the still (m)
H	Mean height of the still (m)
h_L	Latent heat of vaporization ($kJ\ kg^{-1}$)
h_{ev}	Evaporative heat transfer coefficients ($kJ\ kg^{\wedge -1}$)
hc_{ge-amb}	Convective heat transfer coefficient between outer glass cover and ambient ($Wm^{-2}\ K^{-1}$)
hc_{sw-gi}	Convective heat transfer coefficient between saline water and inner glass ($Wm^{-2}\ K^{-1}$)
h_{ge-sky}	Heat transfer coefficient between outer glass cover and sky ($Wm^{-2}\ K^{-1}$)
$\dot{m}_{sw}$	Mass flow rate of Saline water ($kg\ m^{-3}$)
$\dot{m}_b$	Mass flow rate of brine ($kg\ m^{-3}$)
m_w	Mass yield hourly ($kg\ m^{-3}$)
$\dot{m}_{ev}$	Mass rate of produced vapor ($kg\ m^{-3}$)
Nu	Nusselt number (dimensionless)
N	Exponent in Nusselt correlation (dimensionless)
Pr	Prandtl number (dimensionless)
P_{gi}	Partial pressure of the water at the interior of the glass cover (Pa)
P_{sw}	Partial pressure at saline water surface temperature (Pa)
Ra	Rayleigh number (dimensionless)
T_{amb}	Ambient temperature (K)
$T_{b,out}$	Brine output temperature (K)
T_{ge}	Temperature at the outer side of cover glass (K)
T_{gi}	Temperature at the inner side of cover glass (K)
T_{sw}	Temperature of saline water (K)
$T_{sw,in}$	Inlet temperature of saline water (K)

T_{sky}	Sky temperature (K)
t	Time (s)
t_c	Time period for calculation of the yield (s)
U_{loss}	Loss factor per unit surface (W m^{-2} K)
V_g	Volume of glass cover (m^3)
V_{sw}	Volume of saline water (m^3)
v	Wind speed (m s^{-1})
w	Width of the solar stills (m)
x_b	Concentration of salt in brine (mg l^{-1})
x_{sw}	Concentration of salt in feed saline water (mg l^{-1})

Greek letters symbols

$\alpha_c,$	Fraction of solar energy absorbed by basin liner material (dimensionless)
α_g'	Fraction of solar energy absorbed by glass cover material (dimensionless)
α_{sw}'	Fraction of solar energy absorbed by saline water (dimensionless)
β	Coefficient of volumetric thermal expansion (K^{-1})
β_{sw}	Coefficient of volumetric thermal expansion for saline water (K^{-1})
ϵ_{eff}	Effective emissivity (dimensionless)
ϵ_g	Emissivity of cover glass (dimensionless)
ϵ_{sw}	Emissivity of saline water (dimensionless)
φ	Feed recovery rate (dimensionless)
Γ	Yield (kg)
λ	Thermal conductivity (W m^{-1} K^{-1})
λ_g	Thermal conductivity of glass cover (W m^{-1} K^{-1})
λ_{sw}	Thermal conductivity of saline water (W m^{-1} K^{-1})
λ_{in}	Thermal conductivity of insulation material (W m^{-1} K^{-1})
μ	Viscosity (N s m^{-2})
μ_{sw}	Viscosity of saline water (N s m^{-2})
ρ	Density(kg m^{-3})
ρ_g	Glass cover density (kg m^{-3})
ρ_{sw}	Saline water density (kg m^{-3})
σ	Stefan-Boltzmann^'s constant (5.6697 $\times$ 10^{-8}) (W m^{-2} K^{-4})
θ	Dew point temperature (K)
ϵ	Relative humidity (dimensionless)
ΔT	Temperature difference in transfer by natural convention (K)

Appendix A

Parameter	Value
Basin Area	0.2
Thickness	2
Basin Material	Aluminum
Insulation	Wool
Thickness	20
Channels	PVC
Glass	Tempered
Glass Area	0.234
Thickness glass cover	0.04
FRL Lens	R18
Lens Area	0.2839

Table 1.
Material properties of solar stills.

Parameters	Value
Basin absorptivity	0.90
Glass absorptivity	0.05
Water reflectivity	0.05
Glass reflectivity	0.05
Glass emissivity	0.94
Water emissivity	0.95
Specific heat	4.002
Glass thermal conductivity	1.03
Insulation thermal conductivity	0.0363
Water depth	0.02
Wind velocity	4.6

Table 2.
Thermophysical properties.

Cost(PKR)	Without FRL	With FRL
Capital	3550	19500
Operational	1200/year	1200/year
Maintenance	500/year	500/year

Table 3.
Cost of solar stills with and without FRL lens.

Author details

Zafar Abbas*, Nasir Hayat, Anwar Khan and Muhammad Irfan
Department of Mechanical Engineering, University of Engineering and Technology,
Lahore, Pakistan

*Address all correspondence to: zafarabbaskhaplu@gmail.com

IntechOpen

References

[1] Kahlown AMMA. Water-resources situation in Pakistan: Challenges and future strategies. In: Water Resources in the South: Present Scenario and Future Prospects. Islamabad: Islamabad, COMSATS' Series of Publications on Science and Technology; 2003. p. 199

[2] Tiwari HGN. Present status of solar distillation. Desalination. 2003;75: 367-373

[3] Jones E, Qadira M, van Vliet MTH, Smakhtin V, Kang S-m. The State of desalination and brine production: A global outlook. Science of the Total Environment. 2019;657:1343-1356

[4] Malakani MS. A review of coal and water resources of Pakistan. Science, Technology and Development. 2012;31: 202-218

[5] Ministry of Planning and Development. Water and Sanitation. Pakistan: Ministry of Planning and Development; 2017

[6] Asian Productivity Organization. Handbook on Green Productivity. Canada: Asian Productivity Organization; 2006. p. 341

[7] Hamadou KAOA. Modeling an Active Solar Still for Sea Water Desalination Process Optimization. UK: Elsevier; 2014. p. 8

[8] Kucera J. Desalination; Water from Water. Salen Massachusets: Scrivener Publishing LLC; 2014

[9] McAdams W. Heat Transmission. Third ed. Tokyo, Japan: McGraw-Hill Kogakusha; 1954

[10] Dunkle R. Solar water distillation; the roof type still and a multiple effect diffusion still. In: International Developments in Heat Transfer ASME. University of Colorado, Colorado:

Proceeding of International Heat Transfer Part V; 1961

[11] Clark PBE. Radiative Cooling: Resources and Applications. Amherst, MA: Proceedings of the Passive-Cooling Workshop; 1980

[12] Murray F. On the computation of saturation vapor pressure. Journal of Applied Meteorology. 1967;6:203-204

[13] Rahbar JEN. Estimation of Convective heat transfer coefficient in a single-slope solar still: A numerical study. Desalination and Water Treatment. 2012;50:387-396

[14] Ghadhban A. An Analysis and Simulation of Solar Water Desalination Systems. Iraq: University of Basrah; 2017

[15] Malik GTAKMSMAS. Solar Desalination. UK: Pregamon Press Ltd; 1982

Chapter 6

Performance Investigation of the Solar Membrane Distillation Process Using TRNSYS Software

Abdelfatah Marni Sandid, Taieb Nehari, Driss Nehari and Yasser Elhenawy

Abstract

Membrane distillation (MD) is a separation process used for water desalination, which operates at low pressures and feeds temperatures. Air gap membrane distillation (AGMD) is the new MD configuration for desalination where both the hot feed side and the cold permeate side are in indirect contact with the two membrane surfaces. The chapter presents a new approach for the numerical study to investigate various solar thermal systems of the MD process. The various MD solar systems are studied numerically using and including both flat plate collectors (the useful thermal energy reaches 3750 kJ/hr with a total area of 4 m^2) and photovoltaic panels, each one has an area of 1.6 m^2 by using an energy storage battery (12 V, 200 Ah). Therefore, the power load of solar AGMD systems is calculated and compared for the production of 100 L/day of distillate water. It was found that the developed system consumes less energy (1.2 kW) than other systems by percentage reaches 52.64% and with an average distillate water flow reaches 10 kg/h at the feed inlet temperature of AGMD module 52°C. Then, the developed system has been studied using TRNSYS and PVGIS programs on different days during the year in Ain Temouchent weather, Algeria.

Keywords: solar desalination, membrane distillation, photovoltaic system, solar-thermal system, cooling system

1. Introduction

Nowadays, on the world level, the demand for drinking water is in strong growth. In fact, to face the rapid increase in water demand in the irrigation and industrial sectors, as well as in the incompressible needs of the population in the large agglomerations of the various countries, highlights the research in the desalination of water as a capital due to the fact that the water scarcity increases in countries where water resources are too low in relation to population and agriculture [1].

Solar membrane distillation (SMD) was considered an appropriate water provision option in decentralized regions. For rural arid populations with less robust infrastructure availability, it is not cost effective to scale down conventional desalination technologies, such as reverse osmosis (RO) or multistage flash distillation (MSF) [2]. Moreover, MD requires less vapor space and building material quality compared to conventional thermal distillation processes leading to potentially lower construction costs [3]. As mentioned earlier, another reason for coupling membrane distillation technology with solar energy sources is due to its high tolerance to fluctuations in operating conditions and operation with low-grade thermal energy. Since MD is operated at a similar range of temperatures obtained from low-temperature solar heaters, integration is straightforward [4]. The selection of solar collectors depends upon the location, and low-grade thermal energy-absorbing collectors like flat plate (FPC) and evacuated tubular (ETC) collectors have been considered as thermal sources for MD by many researchers. Apart from these, compound parabolic collectors (CPC), solar stills, and solar gradient ponds were also considered as thermal sources for various MD systems [5–7].

Recently, various MD processing technics have appeared as solutions that have a free energy source and diversity of membrane distillation technologies such as direct contact membrane distillation (DCMD), vacuum membrane distillation (DMV), air gap membrane distillation (AGMD), and sweeping gas membrane distillation (SGMD) [8]. According to the high-energy costs associated with existing desalination methods, there is a great demand for technologies that can use low-temperature sources like waste heat or solar energy. DCMD is the most MD config-uration technology studied due to the simplicity and ease of handling, where its energy efficiency, called the membrane thermal efficiency (MTE), is commonly related to the operating conditions [9]. In the MD process field, the DCMD process has a lower MTE against the AGMD procedure because of conduction heat losses. The mechanism functions of the AGMD systems are based on the stagnant air gap interposition between the membrane and condensation area, which leads to an inherently increase in the thermal energy efficiency of the process [10]. Conse-quently, the first patent to discuss the principle of AGMD appeared with Hassler [11] and Weyl [12] for the basics knowledge, in which the concept and behavior of AGMD systems can be found in different literature studies [13–15]. Hanemaaijer et al. [16] introduced an idea of internal heat recovery that is called memstill membrane distillation. Sequentially, Duong et al. [17, 18] conducted a study that allowed only AGMD to restore the latent heat without any external heat exchanger. Minier-Matar et al. [19] found through their study that AGMD provides a higher resistance to mass transfer and runs at low water flow.

Although recent developments in AGMD configurations, the first flat plate AGMD system was developed by the Swedish Svenska Utvecklings AB in 2016 [20], while such modules today have been manufactured and commercialized by Scarab development. Each module is made up of 10 planar cassettes with an overall mem-brane surface of 2.3 m² and a global capacity of 1 to 2 m³/day of distillate water [21]. The single stage consists of injection-molded plastic frames containing two parallel membranes, feed and exit channels for warm water, and two condensing walls [20, 22]. Achmad et al. [23] developed a portable hybrid solar membrane distillation system for the production of freshwater using vacuum multi-effect MD. The total volume distillate output during the test was approximately 70 L with an approxi-mate conductivity of 4.7 µS/cm. The average distillate output rate was 11.53 L/hr. with a maximum of 15.94 L/hr. at noontime, whereas the distillate flux was in the range of 1.5 to 2.6 L/m² hr.

Dynamic simulation of the combined system using tools such as TRNSYS and parametric analysis enables to design of a functional system and then optimizes it. In this study by Kumar et al. [24], the application of the cogeneration system for residential households in the UAE is considered for per capita production of 4 L/day of pure water and 50 L/day of domestic hot water. The optimized cogeneration system utilizes more than 80% of the available solar energy gain and operates between 45 and 60% collector efficiencies for FPC and ETC systems, respectively.

The cogeneration operation reduces 6–16% of thermal energy demand and also enables 25% savings in electrical energy demand.

The configuration studied in this article for seawater desalination is AGMD. It is a thermal membrane process, the driving force behind the transfer being a partial pressure difference on either side of the membrane created by a temperature difference. Compared to other membrane distillation techniques, AGMD seems interesting for its aspects of low membrane wetting and the fact that there is no additional energy consumption linked to the use of an additional pump. In recent years, AGMD has experienced more sustained development, mainly in research, thanks to developments in membrane manufacturing techniques. Seawater desalination is one of the most promising fields for the application of solar thermal energy due to the coincidence, in many places of the world, of water scarcity, seawater availability, and good levels of solar radiation (like Algeria). The solar membrane distillation (SMD) is recently an under-investigation desalination process that is suitable for developing self-sufficient, small-scale applications. The use of solar energy considerably reduces operating costs. The main objective of this project is to analyze and optimize renewable-driven AGMD systems.

This objective can contribute to ensure the availability of distillate water by using the solar desalination process. In this work, a complete test rig to evaluate the performance of the membrane distillation module driven by solar energy during the flat plate and evacuated tube collector heating process is studied throughout Ain Temouchent weather, Algeria. Additionally, this study contains a comparison of different renewable energy systems integrating an air gap membrane distillation module.

2. Materials and methods

2.1 Description of the thermal system

The studied system contains the thermal energy loop, as shown in **Figure 1**. The system incorporating a flat plate collector (FPC) with an area of 4 m^2 providing heat *via* a freshwater heat transfer fluid to a storage tank containing 300 L with heat exchanger internal and auxiliary heaters, a pump, and a controller in differential temperature. **Figure 2** shows the total volume of daily consumption of 100 L/day and the volume of consumption in each hour of the day.

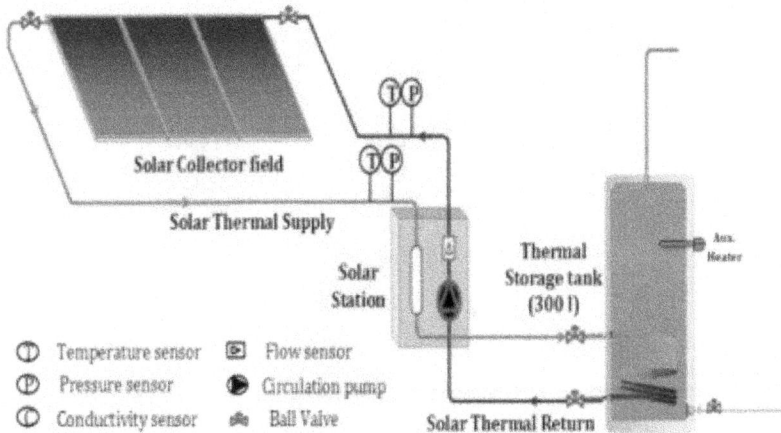

Figure 1.
Diagram of a solar hot water system.

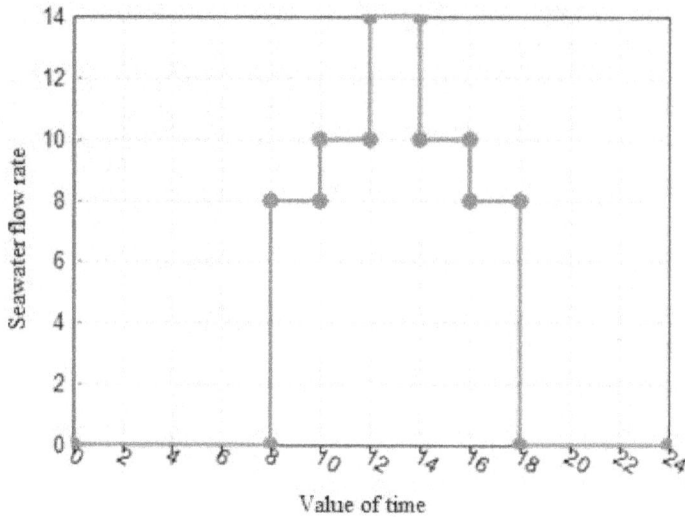

Figure 2.
Daily seawater consumption profile (100 L/day).

2.1.1 The thermal tank storage

The thermal tank is equipped with inlet and outlet pricking and installation of accessories, in all there are 10 prickings, the cold water inlet is at the bottom of the tank at reference A, the exit is at the top of the tank (J), the inlet of the hot fluid coming from the panel is connected to the point C and its exit to the B, in the case of a recirculating installation we use the pricking D, the auxiliary resistance is connected to the E reference, the installation of temperature probes is positioned at references G and H, the thermometer has been connected to the position I, and the point F is reserved for a manhole or a second auxiliary resistance; for more details see **Figure 3**.

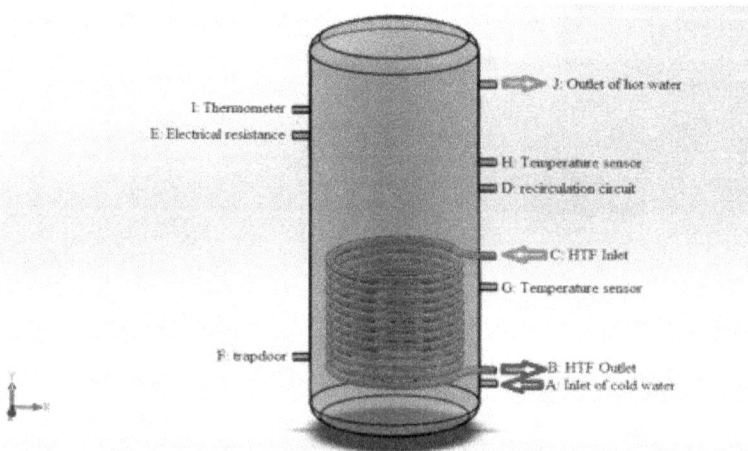

Figure 3.
The thermal storage tank using SolidWorks software.

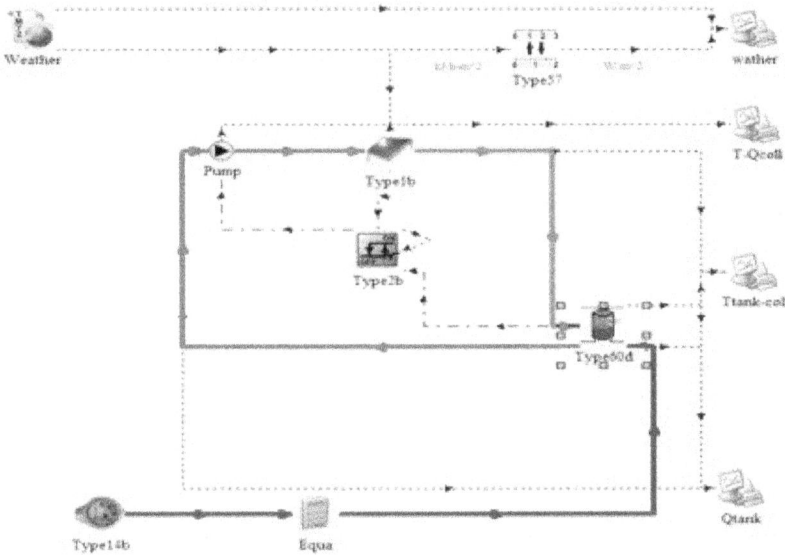

Figure 4.
Assembly diagram of the solar system in the TRNSYS simulation.

The thermal tank is completely covered with a layer of 100-mm-thick glass wool insulation and a mild steel cover sheet of 7/10 mm. A thermovitrification lining should be applied against the corrosions of the product to be heated, and a magnesium anode is installed at the top of the tank.

2.1.2 The thermal system in the TRNSYS software

The model of the solar water heating system is developed by using transient simulation software TRNSYS, as shown in **Figure 4**. The following solar system components are used:

- Reading and processing of meteorological data (TYPE109-TM2)

- Single speed pump (Type 3b)

- Flat plate collector (Type 1b)

- Storage tank with optional internal auxiliary heaters and optional internal heat exchangers with 1 input and 1 output (Type 60d)

- Distribution Water Supply Profile (Type 14e)

- Differential temperature controller (Type 2b)

- Online plotters with files (Type 65c)

The parameters of the solar system components of this model appear in **Tables 1–3**.

Parameters	Value	Unit
Tank volume	300	l
Tank height	1.42	m
Height of flow inlet 1	0.215	m
Height of flow outlet 1	1.415	m
Fluid specific heat	3.911	kJ/kg k
Maximum heating rate	0	kW
Heat exchanger inside diameter	0.02	m
Heat exchanger outside diameter	0.027	m
Heat exchanger fin diameter	0.027	m
Total surface area of heat exchanger	1.2	m^2
Heat exchanger length	18	m
Height of heat exchanger inlet	0.805	m
Height of heat exchanger outlet	0.215	m

Table 1.
Hot water cylinder.

Parameters	Value	Unit
Number in series	2	—
Collector absorber area	2	m^2
Intercept efficiency	0.778	kJ/kg k
inlet flow rate	20	kg/hr

Table 2.
Solar collector parameters.

Parameters	Value	Unit
Rated flow rate	20	kg/hr
Rated power	0.4	kW

Table 3.
Pump parameters.

2.2 The cooling system

Desalination using distillation membranes requires a very accurate study of all aspects, and because the cooling system has a significant role in the production of distilled water, in this research we study the numerical model of cooling the membrane with the use of the heat exchanger. Therefore, the power load of the heat exchanger is calculated, and the cooling system is studied with the help of the TRNSYS program in Ain Temouchent weather, Algeria.

2.2.1 Description of the cooling system

In the distillation membrane cooling unit, the distilled water tank is filled with a least half a liter and is circulated by a pump to be passed to the distillation membrane. On the opposite side, water vapor produces from the hot system. Thus, when

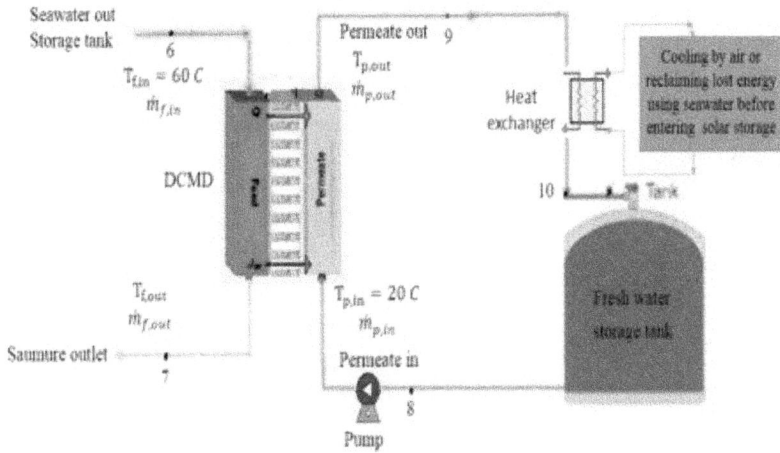

Figure 5.
Cooling system for distillation membrane.

water vapor is increased, the volume of distilled water is increased. The role of the heat exchanger is to reduce the temperature of the distilled water mixed with steam that is 50–60°C to 20–25°C. This makes the mixture state liquid "distilled water" as shown in **Figure 5**.

2.2.2 Cooling system in the TRNSYS software

A unit for cooling the distillation membrane is created at the program TRNSYS as shown in **Figure 6** with the following tools:

- Reading and processing of meteorological data (TYPE109-TM2)

- Single speed pump (Type 3b)

- Storage tank with 1 input and 1 output (Type 60d)

- The counterflow heat exchanger (Type 5b)

- Online plotters with files (Type 65c)

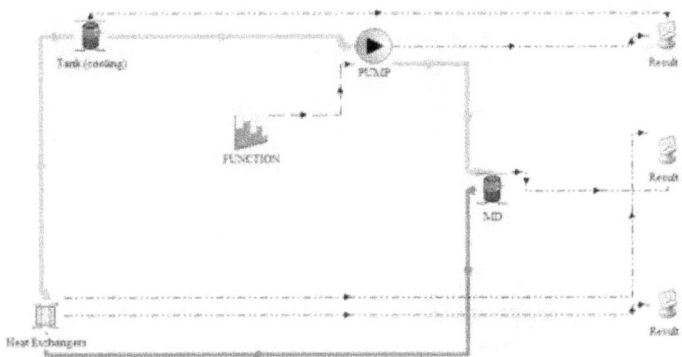

Figure 6.
Assembly diagram of the solar system in the TRNSYS simulation.

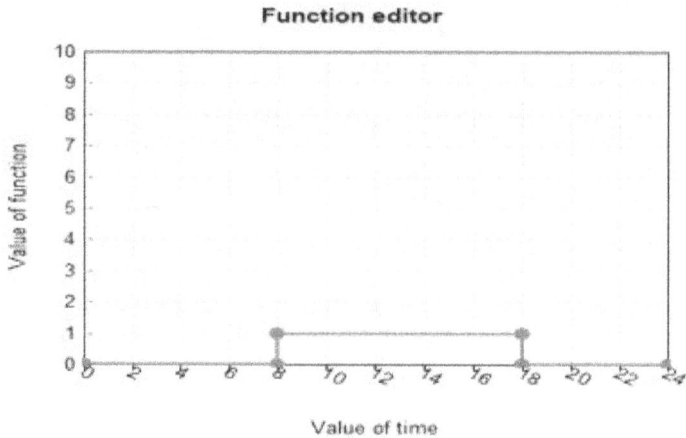

Figure 7.
Value of time to operate cooling unit for distillation membrane.

This system used to cool the distillation membrane relies on the heat exchanger in particular because it reduces the temperature coming out of the membrane, and thus be there more flow for distillated water and increases the production.

In order to obtain good harmonic results, we chose the time from 8 AM to 6 PM to operate the distillation membrane-cooling unit as shown in **Figure 7**. This the time when human needs in it distilled water in many fields.

2.3 Photovoltaic (PV) system

Solar PV system includes different components that should be selected according to the system type, site location, and applications. The major components for solar PV system are solar charge controller, inverter, and battery bank.

To save costs, a photovoltaic system uses based on renewable energy (solar energy). Therefore, the energy needed calculates for the pumps and replaces by photovoltaic panels, each one has an area of 1.6 m^2 by using an energy storage battery (12 V, 200 Ah). **Figure 8** presents a photovoltaic system with energy storage.

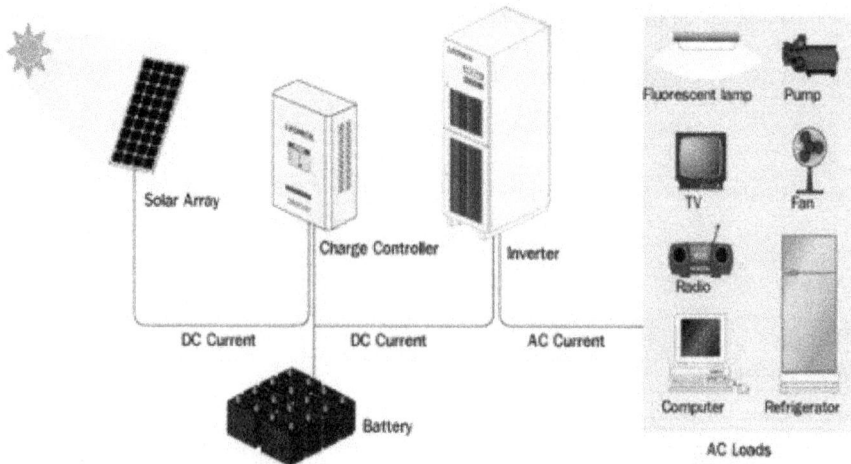

Figure 8.
Synoptic representation of the structure of a photovoltaic system with energy storage.

2.4 The AGMD module

AGMD is a configuration of membrane distillation (MD) in which an air layer is interposed between a porous hydrophobic membrane and the condensation surface. The module contains the cassette in a plate and frame configuration and the layout of components in the bench-scale MD module. The cassette of the AGMD module has the following specifications [24]—hydrophobic PTFE membrane with a pore size of 0.2 µm, the thickness of 280 µm, and total membrane area of 0.2 m².

3. Equations and methods

The most significant influential design variables on the AGMD performance are the feed inlet temperature (T_{Hin}), the cooling inlet temperature (T_{Cin}), which is condensation temperature, the feed flow rate (V_f), and feed concentration (C_f). The selected performance indicators of the AGMD process are distillate flux (D_w) and specific performance ratio (SPR), whereas D_w is calculated by [24]:

$$D_w = \frac{M_d}{St} \tag{1}$$

where M_d (kg) is the mass of distillate water collected within the time t and S (m²) is the effective membrane surface area of evaporation. SPR is obtained by [20]:

$$SPR = \frac{M_d}{Q_{md}} \tag{2}$$

Q_{md} (KWh) is the thermal energy supplied to the AGMD module.

The regression quadratic model with coded parameters [24] can be expressed as follows:

$$Y = \beta_0 + \beta_1 X_1 + \beta_2 X_2 + \beta_3 X_3 + \beta_{12} X_1 X_2 + \beta_{13} X_1 X_3 + \beta_{23} X_2 X_3 + \beta_{11} X_1^2 + \beta_{22} X_2^2 + \beta_{33} X_3^2 \tag{3}$$

Kumar et al. [24] determined the final regression equations for D_w and T_{Hout} in terms of actual operating parameters as follows:

$$D_w = -6.57 + 0.16 \times T_{Cin} + 0.15 \times T_{Hin} - 5.86 \times 10^{-3} \times V_f - 5.77 \times 10^{-3} \\ \times T_{Cin} T_{Hin} - 2.5 \times 10^{-4} \times T_{Cin} V_f + 3.44 \times 10^{-4} \times T_{Hin} V_f + 2.48 \times 10^{-3} \\ \times T_{Hin}^2 \tag{4}$$

$$T_{Hout} = 3.097 + 6.82 \times 10^{-2} \times T_{Cin} + 0.772 \times T_{Hin} + 3.5 \times 10^{-3} \times V_f + 1.42 \times 10^{-3} \\ \times T_{Cin} T_{Hin} \tag{5}$$

The basic method of measuring collector performance is to expose the operating collector to solar radiation and measure the fluid inlet and outlet temperatures and the fluid flow rate. The useful gain is

$$\dot{Q}_u = m_0 C_{pf} (T_0 - T_i) \tag{6}$$

m_0 is the solar fluid mass flow rate (kg/hr), C_{pf} is the specific heat capacity of solar fluid (kJ/hr), and T_0 and T_i are the inlet and outlet temperatures of the solar fluid (K). The efficiency of flat plate collectors is expressed as follows [5]:

$$\eta = \eta_0 - a_0 \times \frac{(T - T_{amb})}{G} - a_1 \times \frac{(T - T_a)2}{G} \tag{7}$$

With G the solar flux and T_a the ambient temperature, a_0 and a_1 (W/m^2 K^2) are characteristic constants of the efficiency of the collector.

Heat exchanger counterflow effectiveness is [5]:

$$\varepsilon = \frac{1 - \exp\left(-\frac{UA}{C_{min}}\left(1 - \frac{C_{min}}{C_{max}}\right)\right)}{1 - \left(\frac{C_{min}}{C_{max}}\right)\exp\left(-\frac{UA}{C_{min}}\left(1 - \frac{C_{min}}{C_{max}}\right)\right)} \tag{8}$$

UA is the overall loss coefficient between the heater and its surroundings during operation (kg/hr), C_{max} is the maximum capacity rate (kJ/hr. K), and C_{min} is the minimum capacity rate (kJ/hr. K).

Required heating rate including efficiency effects in the auxiliary heaters is [6]:

$$Q_{aux} = Q_{loss} + Q_{fluid} \tag{9}$$

With:

$$Q_{loss} = U A \left(\overline{T} - T_{env}\right) + (1 - \eta_{htr})Q_{max} \text{ and } Q_{fluid} = \dot{m}_0 C_{pf} \left(T_{set} - T_i\right)$$

Q_{aux} is the required heating rate including efficiency effects (kg/hr), Q_{fluid} is the rate of heat addition to fluid stream (kg/hr), Q_{loss} is the rate of thermal losses from the heater to the environment (kg/hr), Q_{max} is the maximum heating rate of the heater (kg/hr), η_{htr} is an efficiency of the auxiliary heater, m_0 is the outlet fluid mass flow rate (kg/hr1), C_{pf} is the fluid specific heat (kJ/hr), T_i is the fluid inlet temperature (K), $(\overline{T})$ is the brackish water average temperature, T_{set} is the set temperature of heater internal thermostat (K), and T_{env} is the temperature of heater surroundings for loss calculations (K).

The PV system is calculated the following equations [7]:

The peak power of the autonomous photovoltaic installation is

$$P_c = P_{pv} = \frac{D}{N \times F} \tag{10}$$

Pc Power of the PV field, D Daily need kWh/day, F Form factor, N Number of hours equivalent.

$$N = \frac{G_T(t)}{G_{T,STC}} \tag{11}$$

$G_T(t)$ is the solar radiation incident on the solar PV array in the current time step kW/m^2; $G_{T,STC}$ is the incident radiation at standard test conditions kW/m^2.

4. The solar AGMD system

The model of the solar thermal AGMD system is developed by using TRNSYS software, which is a quasi-steady-state simulation program. TRNSYS is a transient

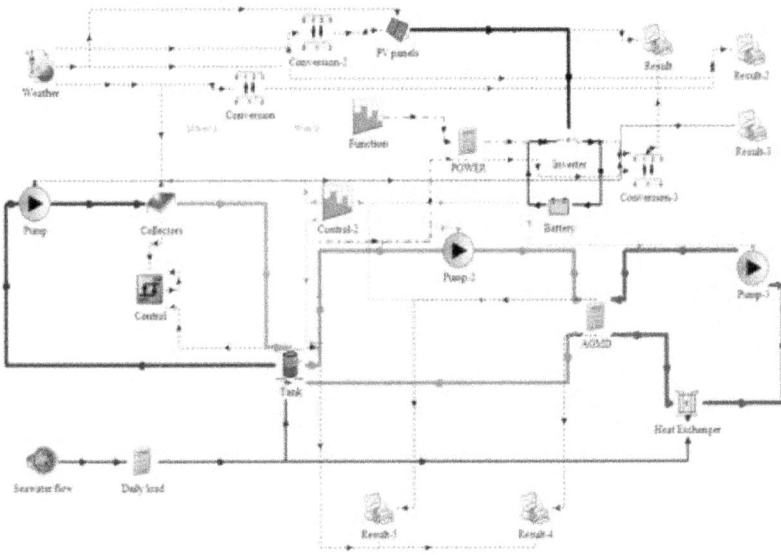

Figure 9.
Assembly diagram of the AGMD system in the TRNSYS simulation.

systems simulation program. A TRNSYS simulation project consists in choosing a set of mathematical models of physical components (by relying either on existing models in the TRNSYS model libraries or by creating them) and in describing the interactions between these models. TRNSYS contains a large number of standard models (utilities, thermal storage, equipment, loads and structures, heat exchangers, hydraulics, regulators, electrical/photovoltaic components, solar collectors).

The main component of the model is the AGMD unit, which is represented by a new equation in TRNSYS. As shown in **Figure 9**, additional components to the model include TYPE109-TM2 reading and processing of meteorological data, Type 91 heat exchanger, Type 60 storage tank, Type 1 flat plate collector, Type 2 differential temperature controller, Type 3 single speed pump, Type 6 auxiliary heaters, Type 94 photovoltaic panels, Type 47 storage battery, Type 48 inverter, Type 14 forcing functions, Type 57 unit conversion, and Type 65 online plotter. **Tables 1–3** show the values of parameters that are used in the TRNSYS model.

5. Results and discussion

5.1 Weather data

This model is installed in Ain-Temouchent weather (latitude 35° 3′0″N, longitude 1° 1′0″E in Algeria). **Figure 10** shows changing climate conditions throughout the year:

The weather of the state Ain-Temouchent is pleasant, warm, and moderate in general. At an average temperature of 25.7°C, August is the hottest month of the year. At 10.8°C on average, January is the coldest month of the year. Therefore, in this **Figure 10a**, we notice a change in temperature throughout the year, which reaches up to 40°C in the month of August, and we note that the wind is fairly moderate and does not exceed the speed of this one 15 m/s. For the irradiation, it

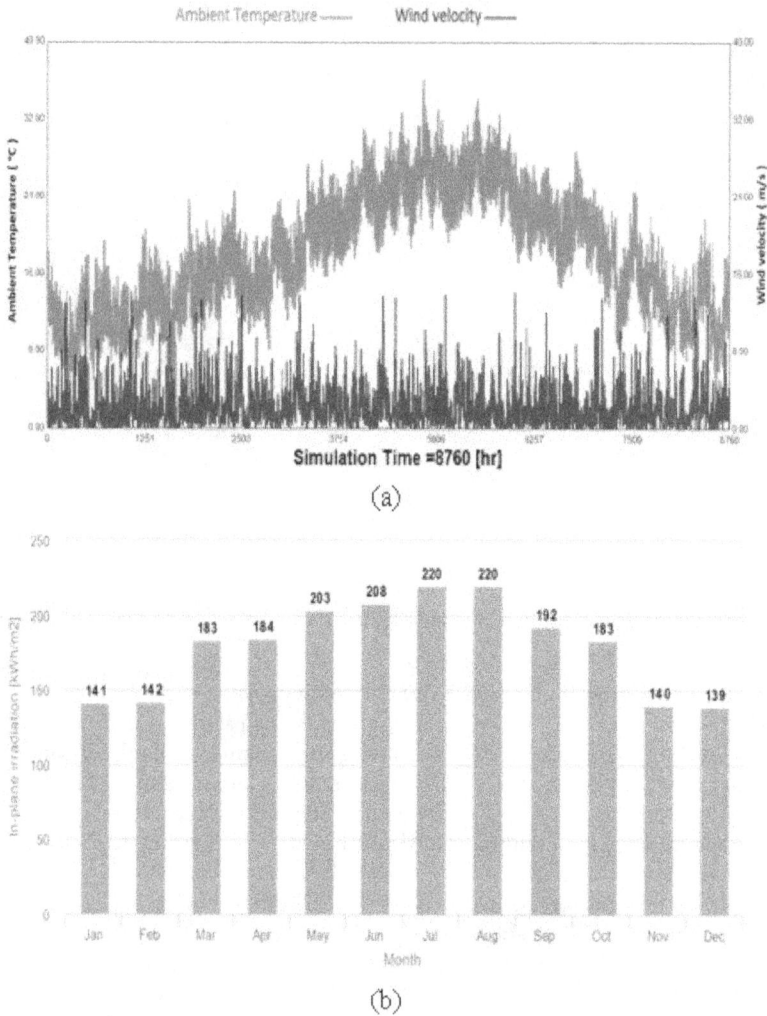

Figure 10.
(a) Ambient temperature, wind speed, and (b) irradiation in the Ain-Temouchent weather.

changes during the months of the year and reaches up to 220 kWh/m^2 in the month of August and July, as shown in **Figure 10b**, because the temperature is high in this period of the year.

5.2 The outlet temperatures and the useful thermal energy Qu for FPC system

The change of outlet temperatures of collectors and storage tank for the thermal system without using an auxiliary heater is illustrated in **Figure 11** on day 1st of January and August. The results show that the temperature decreases in January that reaches between 51 and 79°C (tank storage and collectors), but it increases in August when the temperature is high and reaches between 59 and 87°C, respectively. This change is due to the change in ambient temperature and radiation in the daytime and their difference from month to month as shown in **Figure 10**.

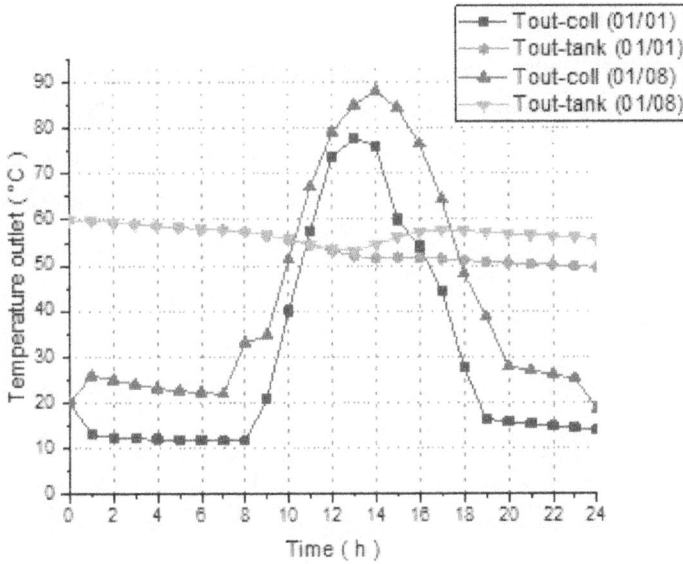

Figure 11.
The outlet temperatures of the solar energy collector and hot water outlet tank.

Figure 12 illustrates the amount of useful thermal energy obtained from collectors in different climatic conditions in January and August. We note that the productivity of the useful thermal energy is low in January and reaches 3250 kJ/hr. and increases in August, which reaches 3750 kJ/hr. Consequently, the changes in temperature and climatic conditions as clarified in **Figure 10** affect the useful thermal energy in different months selected. Therefore, when the temperature decreases, the useful thermal energy also decreases and vice versa.

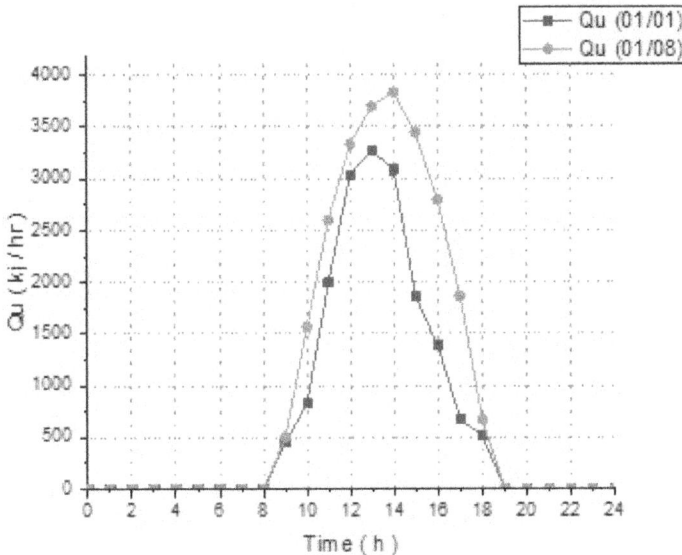

Figure 12.
The useful thermal energy Qu for FPC system.

5.3 The outlet temperatures in the storage tank

Figure 13 shows the temperature at the outlet of the storage tank at 60°C. In the storage tank, the initial temperature of the specified nodes into the tank's stratification (T1, T2, T3, T4, and T_{out} system) coming out from the heat exchanger in the tank has been calculated. They reach a maximum of 37°C on this first day of the cold month of January. For this reason, the auxiliary heater should be added to reach the temperature at 60°C when the weather temperature is low.

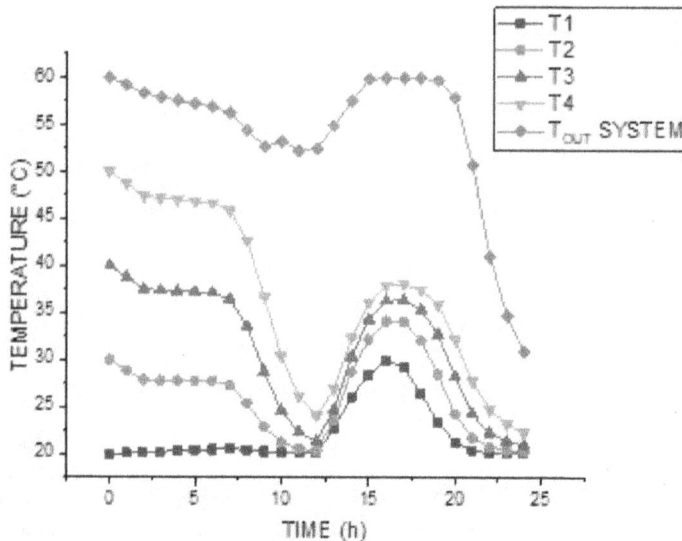

Figure 13.
The outlet temperatures in the storage tank for the FPC system.

5.4 Outlet cold and hot temperatures of the cooling system

Table 4 represents the changes of outlet cold and hot temperatures coming out of the heat exchanger (HX) at the heat transfer coefficient of the counter flow HX using the inlet temperatures between 20 and 50°C, respectively.

U A (kJ/hr. k) "Heat Exchanger"	Outlet hot temperature (°C)	Outlet cold temperature (°C)
100	28.87	41.16
150	26.56	43.47
200	25.20	44.82
250	24.31	45.72
300	23.68	46.35
350	23.21	46.82
400	22.85	47.18
500	22.32	47.70

U A, overall heat transfer coefficient of the counterflow heat exchanger.

Table 4.
Outlet cold and hot temperatures of the heat exchanger with (U A).

In order to improve the distillation membrane-cooling unit, the temperature changes and the overall heat transfer coefficient of the counter flow heat exchanger were calculated. We note that increasing the overall heat transfer coefficient increases the temperature on the cold side of the heat exchanger by 20 to 47.70°C. In addition, using the overall heat transfer coefficient (400–500 kJ/hr) is obtained the maximum production of distilled water in the membrane distillation process.

In **Figure 14**, the temperature coming from the distillation membrane is the same as the temperature entering the heat exchanger. This curve shows that the temperature of the distillation membrane reduces from 50 to 22.32°C in the hot side and increases from 20 to 47.7°C in the cold side. Therefore, the temperature of the hot side from the heat exchanger is reverse on the cold side.

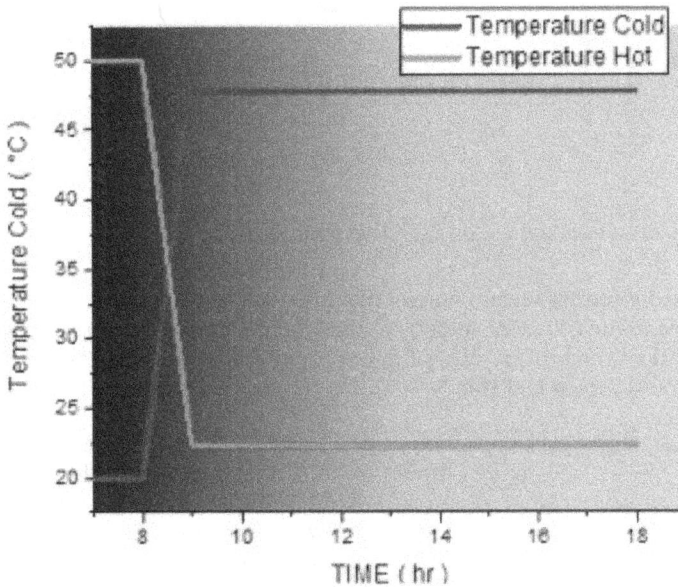

Figure 14.
Outlet cold and hot temperatures of the cooling system.

5.5 The inlet temperatures and distillate water flows of the AGMD module

Figure 15 presents the inlet temperatures and distillate water flows of the AGMD module on the day of 1st January. The MD solar systems are studied numerically using and including both flat plate collectors and photovoltaic panels. Therefore, the power load of solar AGMD systems is calculated and compared for the production of 100 L/day of distillate water. It was found that the developed system obtains an average distillate water flow that reaches 10 kg/hr. at the feed inlet temperature of AGMD module 52°C.

5.6 The daily power consumption and variations of energy of the PV panels

To save costs, a photovoltaic system is used based on renewable energy (solar energy). Therefore, the energy needed calculates for the pumps and replaces by four photovoltaic panels, each one has an area of 1.6 m² by using a TRNSYS help program.

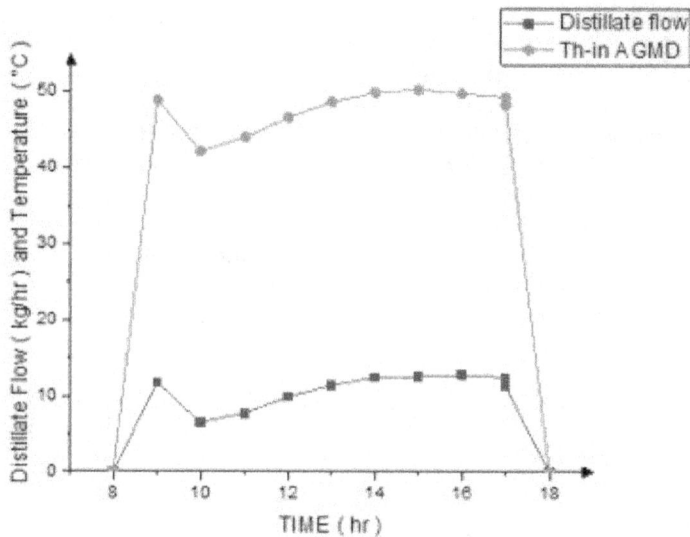

Figure 15.
The inlet temperatures and distillate water flows of the AGMD module.

Estimated electricity requirements (kWh/day): The daily power consumption [or daily requirement in (kWh/day)] is given by the product of the nominal power of the load (kW) and the number of hours of daily use (hr/day). The daily power consumption is shown in **Table 5**.

Devices	Number	Unit power	Frequency or duration of daily use	Power	Energy
Pump of water	1	0.1 kW	10 h	0.1 kW	1 kWh
Pump of collectors	1	0.1 kW	10 h	0.1 kW	1 kWh
Pump of AGMD	1	0.1 kW	10 h	0.1 kW	1 kWh
Pump of cooling system	1	0.1 kW	10 h	0.1 kW	1 Kwh
Heat exchanger of cooling system	1	0.37 kW	10 h	0.37 kW	3.7 kWh
Total				0.77 kW	7.7 kWh

Table 5.
The daily power consumption (water consumption profile 100 L/day).

As shown in **Table 6**, increasing the daily power consumption increased the number of PV panels for AGMD process. Therefore, the various solar thermal systems of the MD process should be investigated to find the developed system that consumes less energy and with an average distillate water flow reaches 10 kg/hr. at the feed inlet temperature of AGMD module 52°C.

5.7 The various solar thermal systems of the MD process

Figure 16 presents a new approach for a numerical study to investigate the various solar thermal systems of the MD process: (1) tank storage with auxiliary heaters, (2) tank storage with two auxiliary heaters, (3) tank storage with a heat exchanger and without auxiliary heaters, and (4) tank storage with a heat exchanger and auxiliary

Total energy	PV panels
10.1 kWh	7 Panels
9.1 kWh	6 Panels
8.1 kWh	5 Panels
7.7 kWh	4 Panels

Table 6.
Variations of energy of the PV panels.

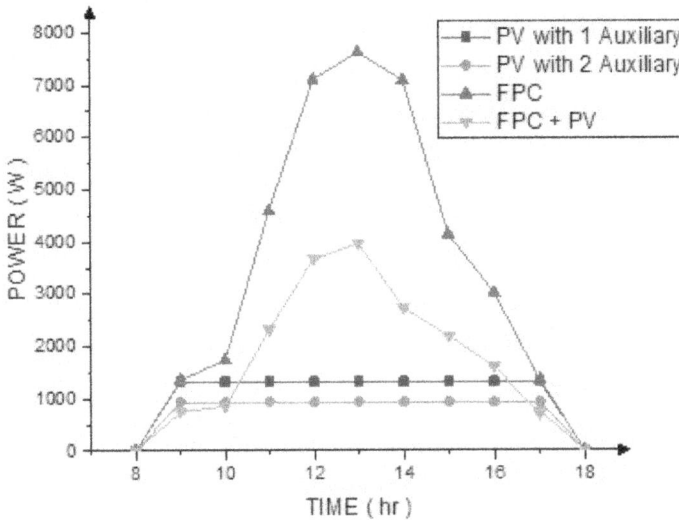

Figure 16.
Power to load of the auxiliary heaters of the thermal system for AGMD.

heaters. The various MD solar systems are studied numerically using and including both flat plate collectors and photovoltaic panels. Therefore, the power load of solar AGMD systems is calculated and compared for the production of 100 L/day of distillate water. It was found that the developed system in case (2) consumes less energy (1.2 kW) than other systems by percentage reaches 52.64% and with an average distillate water flow reaches 10 kg/hr. at the feed inlet temperature of AGMD module 52°C. Then, the developed system has been studied using TRNSYS and PVGIS programs on different days during the year in Ain Temouchent weather, Algeria.

6. Conclusion

A complete test rig to evaluate the performance of the membrane distillation module is driven by solar energy during the flat plate collectors heating process is simulated in Ain-Temouchent, Algeria.

The chapter presents a new approach for a numerical study to investigate the various solar thermal systems of the MD process: (1) tank storage with an auxiliary heater, (2) tank storage with two auxiliary heaters, (3) tank storage with a heat exchanger and without auxiliary heaters, and (4) tank storage with a heat exchanger and an auxiliary heater. The various MD solar systems are studied numerically using and including both flat plate collectors and photovoltaic panels.

Therefore, the power load of solar AGMD systems is calculated and compared for the production of 100 L/day of distillate water. It was found that the developed system in case (2) consumes less energy (1.2 kW) than other systems by percentage reach 52.64% and with an average distillate water flow reaches 10 kg/hr. at the feed inlet temperature of AGMD module 52°C. Then, the developed system has been studied using TRNSYS and PVGIS programs on different days during the year in Ain Temouchent weather, Algeria.

The simulation results show that a very simple AGMD system with a total collectors area of 4 m^2 for the production of 10 L/hr. of distilled water flow throughout the entire year. To save costs, a photovoltaic system is used depending on renewable energy (solar energy). Therefore, the energy needed is calculated for the pumps and is replaced by 4 photovoltaic panels, and each one has an area of 1.6 m^2 using an energy storage battery (12 V, 200 Ah) via TRNSYS and PVGIS to help programs. Accordingly, the purpose of this study is the use of solar panels in the photovoltaic system to produce the necessary electrical energy. The AGMD system using solar energy for seawater desalination will be useful for further simulations or applications of the technology. Finally, these systems of projects that integrate renewable energy technologies with additional services are in principle attractive in terms of the associated socioeconomic benefits.

Acknowledgements

The authors acknowledge financial supports of the FNRSDT/DGRSDT within the framework of ERANETMED3 (Project.ERANETMED3-166 EXTRASEA) from the directorate general for scientific research and technological development.

Author details

Abdelfatah Marni Sandid[1*], Taieb Nehari[1], Driss Nehari[2] and Yasser Elhenawy[3]

1 Smart Structures Laboratory (SSL), University Ain Temouchent Belhadj Bouchaib, Ain-Temouchent, Algeria

2 Department of Mechanical Engineering, University Ain Temouchent Belhadj Bouchaib, Ain-Temouchent, Algeria

3 Department of Mechanical Power Engineering, Port Said University, Port Said, Egypt

*Address all correspondence to: abdelfatahsandid@hotmail.com

IntechOpen

References

[1] Alkhudhiri A, Darwish N, Hilal N. Membrane distillation: A comprehensive review. Desalination. 2012;**287**:2-18. DOI: 10.1016/j. desal.2011.08.027

[2] Koschikowski J, Wieghaus M, Rommel M, Ortin S, Suarez P, Betancort-Rodríguez J. Experimental investigations on solar driven stand-alone membrane distillation systems for remote areas. Desalination. 2009;**248**: 125-131. DOI: 10.1016/j. desal.2008.05.047

[3] Zhani K, Zarzouma K, Ben-Bachaa H, Koschikowskic J, Pfeifle D. Autonomous solar powered membrane distillation systems: State of the art. Desalination and Water Treatment. 2016;**57**: 23038-23051. DOI: 10.1080/ 19443994.2015.1117821

[4] Chen YH, Li YW, Chang H. Optimal design and control of solar driven air gap membrane distillation desalination systems. Applied Energy. 2012;**100**: 193-204. DOI: 10.1016/j. apenergy.2012.03.003

[5] Marni-Sandid A, Bassyouni M, Nehari D, Elhenawy Y. Experimental and simulation study of multichannel air gap membrane distillation process with two types of solar collectors. Energy Conversion and Management. 2021;**243**: 1-14. DOI: 10.1016/j. enconman.2021.114431

[6] Marni-Sandid A, Nehari T, Nehari D. Simulation study of an air-gap membrane distillation system for seawater desalination using solar energy. Desalination and Water Treatment. 2021;**229**:40-51. DOI: 10.5004/dwt.2021.27394

[7] Marni-Sandid A, Nehari D, Elmeriah A, Remlaoui A. Dynamic simulation of an air-gap membrane distillation (AGMD) process using

photovoltaic panels system and flat plate collectors. Journal of Thermal Engineering. 2021;**7**:117-133. DOI: 10.18186/thermal.870383

[8] Khayet M, Matsuura T. Introduction to Membrane Distillation. Book of Membrane Distillation; 2011. 1-16 p. DOI:10.1016/b978-0-444-53126-1.10001-6

[9] Ruh U, Majeda K, Richard J, James T, McLeskey J, Mohammad A, et al. Energy efficiency of direct contact membrane distillation. Desalination. 2018;**433**:56-67. DOI: 10.1016/j. desal.2018.01.025

[10] Qtaishata M, Matsuura T, Kruczek B, Khayet M. Heat and mass transfer analysis in direct contact membrane distillation. Desalination. 2008;219:272-292. DIO:10.1016/j. desal.2007.05.019

[11] Hassler GL. U.S. patent US3129146A [14 April 1964]

[12] Weyl PK. U.S. patent US3340186A [5 September 1967]

[13] Jonsson S, Wimmerstedt R, Harrysson C. Membrane distillation: A theoretical study of evaporation through microporous membranes. Desalination. 1985;**56**:237. DOI: 10.1016/0011-9164 (85)85028-1

[14] Gostoli C, Sarti GC, Matulli S. Low temperature distillation through hydrophobic membranes. Journal of Separation and Purification Technology. 1987;22:855. DOI: 10.1080/ 01496398708068986

[15] Banat FA, Al-Rub FA, Jumah R, Shannag M. Theoretical investigation of membrane distillation role in breaking the formic acid-water azeotropic point: Comparison between Fickian and Stefan-Maxwell-based models.

International Communications in Heat and Mass Transfer. 1999;**26**(6):879-888. DOI: 10.1016/s0735-1933(99)00076-7

[16] Hanemaaijer JH, van Medevoort J, Jansen AE, Dotremont C, Van-Sonsbeek E, Yuan T, et al. Memstill membrane distillation – A future desalination technology. Desalination. 2006;**199**: 175-176. DOI: 10.1016/j. desal.2006.03.163

[17] Vandita T, Shahu B, Thombre S. Air gap membrane distillation: A review. Journal of Renewable and Sustainable Energy. 2019;**11**:45901. DOI: 10.1063/ 1.5063766

[18] Duong HC, Cooper P, Nelemans B, Cath TY, Nghiem LD. Evaluating energy consumption of air gap membrane distillation for seawater desalination at pilot scale level. Journal of Separation and Purification Technology. 2016;**166**: 55. DOI: 10.1016/j.seppur.2016.04.014

[19] Minier-Matar J, Hussain A, Janson A, Benyahia F, Adham S. Field evaluation of membrane distillation technologies for desalination of highly saline brines. Desalination. 2014;**351**: 101-108. DOI: 10.1016/j. desal.2014.07.027

[20] Swaminathana J, Chunga HW, Warsingera DM, AlMarzooqib FA, Arafatb HA. Energy efficiency of permeate gap and novel conductive gap membrane distillation. Journal of Membrane Science. 2016;**502**:171-178. DOI: 10.1016/j.memsci.2015.12.017

[21] Alklaibi AM, Lior N. Membrane-distillation desalination: Status and potential. Desalination. 2005;**171**(2): 111-131. DOI: 10.1016/j. desal.2004.03.024

[22] Camacho L, Dumée L, Zhang J, Li J, Duke M, Gomez J, et al. Advances in membrane distillation for water desalination and purification

applications. Water. 2013;**5**(1):94-196. DOI: 10.3390/w5010094

[23] Chafidz A, Esa DK, Irfan W, Yasir K, Abdelhamid A, Saeed A. Design and fabrication of a portable and hybrid solar-powered membrane distillation system. Journal of Cleaner Production. 2016;**133**:631-647. DOI: 10.1016/j. jclepro.2016.05.127

[24] Uday KN, Martin A. Experimental modeling of an air-gap membrane distillation module and simulation of a solar thermal integrated system for water purification. Desalination and Water Treatment. 2017;**84**:123-134. DOI: 10.5004/dwt.2017.21201

Section 2

Reactive Distillation

Chapter 7

Reactive Distillation Applied to Biodiesel Production by Esterification: Simulation Studies

Guilherme Machado, Marcelo Castier, Monique dos Santos,
Fábio Nishiyama, Donato Aranda, Lúcio Cardozo-Filho,
Vladimir Cabral and Vilmar Steffen

Abstract

Reactive distillation is an operation that combines chemical reaction and separation in a single equipment, presenting various technical and economic benefits. In this chapter, an introduction to the reactive distillation process applied to the biodiesel industry was developed and complemented by case studies regarding the production of biodiesel through esterification a low-cost acid feedstock (corn distillers oil) and valorization of by-products (glycerol) through ketalization. The kinetic parameters of both reactions were estimated with an algorithm that performs the minimization of the quadratic differences between experimental and calculated data through a Nelder-Mead simplex method. A 4th order Runge Kutta method was employed to integrate the conversion or concentration equations used to describe the kinetics of the reactions in a batch reactor. Both processes were simulated in the commercial software Aspen Plus with the estimated kinetic parameters. The results obtained are promising and indicate that the productivity of both processes can be improved with the application of reactive distillation technologies. The simulated esterification process with an optimized column resulted in a fatty acids conversion increase of 84% in comparison to the values lower than 50% obtained in the experimental tests. Solketal production through ketalization also achieved a high glycerol conversion superior to 98%.

Keywords: reactive distillation, biodiesel, esterification, transesterification, simulation

1. Introduction

Reactive distillation is an operation that incorporates chemical reaction and physical separation in a single unit [1]. This process, when applicable, has several potential advantages for the industry when compared to conventional systems, such as the reduction of capital costs, improvement of component separations, use of fewer instruments for monitoring, reduction of energy costs, and improvement in reaction selectivity [2–6]. Among the possible applications of reactive distillation, the separation of azeotropic mixtures and compounds with similar boiling points can be carried out by adding a reagent that promotes the consumption of one component of the mixture and forms a product with markedly different physical properties, thereby favoring separation. However, the use of reactive distillation is more frequent in systems where product formation is limited by chemical equilibrium. This separation technique allows for the constant removal of one or more products promotes or increases reactant conversion [7].

The first patents for the reactive distillation process were published in the 1920s, developed by Backhaus for the production of esters [8–10]. However, few industrial applications were developed before the 1980s [11], when the commercial process for methyl acetate synthesis via reactive distillation with a homogeneous catalyst was patented by Agreda and Partin [12], in collaboration with the Eastman Chemical Company. This application of reactive distillation is considered an exemplary case because of the substantial reduction in process costs (~80%) [13] achieved through the elimination of units, such as reactors and separation columns, and the possibility of heat integration. The conventional methyl acetate synthesis process (**Figure 1**), which comprises 11 different steps and 28 pieces of equipment, was replaced by only a highly integrated reactive distillation column (**Figure 2**), enabling the aforementioned reduction in process costs. Recent uses of reactive distillation in chemicals production such as acetic acid and methanol [14], alkyl carbonates [15], butadiene [16], butyl acetate [17], carboxylic acids and ether [18] and carboxylic esters [19] are described in patents.

Reactive separation conveniently combines the production and removal of one or more products, enabling improvements in reaction productivity and selectivity. However, despite the benefits of this process, the planning and control of reactive distillation columns are hindered by complex interactions between reaction and separation. Operation conditions resulting in the formation of equilibrium between liquid–liquid–vapor phases, low mass transfer rates between liquid and vapor phases or diffusion inside the catalyst (for heterogeneous reactions) and chemical kinetics with reduced reaction rate need to be avoided or minimized [20]. Additionally, since both separation and reaction take place simultaneously in the same unit, the temperatures and pressures required for the two steps must have similar values [21]. If the overlap of operating conditions is not significant, the use of reactive distillation is not recommended [2].

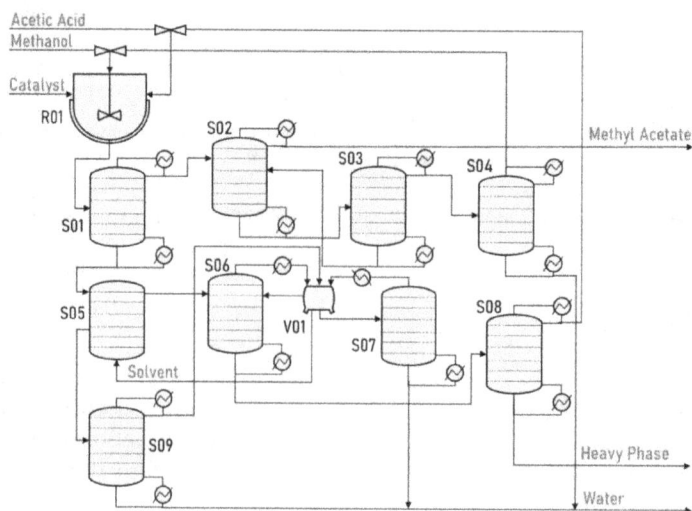

Figure 1.
Schematic representation of the conventional process for the synthesis of methyl acetate. Caption: R01: reactor; S01: mixer; S02: extractive distillation; S03: solvent recovery; S04: methanol recovery (MeOH); S05: extractor; S06: azeotropic column; S07-S09: flash columns; S08: acetic acid recovery; V01: decanter.

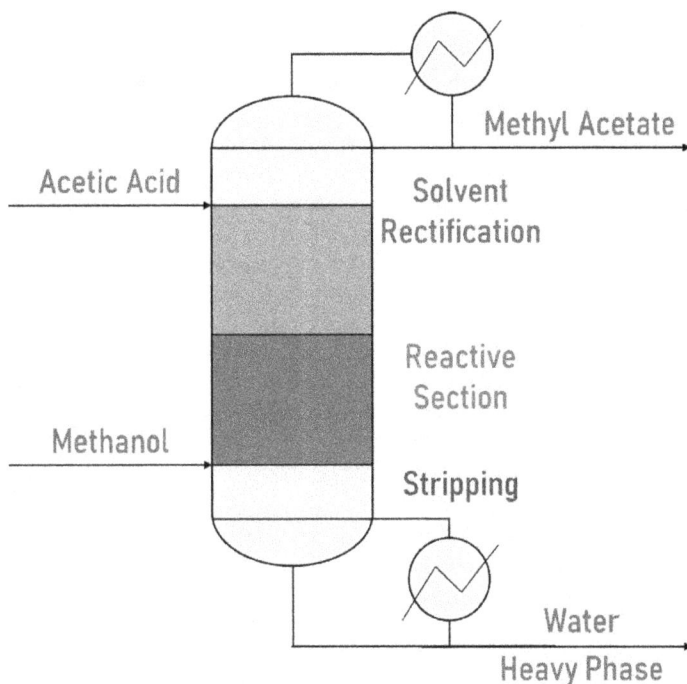

Figure 2.
Schematic representation of the integrated process for the production of methyl acetate by reactive distillation.

2. Reactive distillation - biodiesel applications

The trend toward the use of biofuels has resulted from increased attention to topics related to mitigating environmental impacts by reducing the consumption of fossil fuels [22]. In this context, biodiesel is considered the main substitute fuel for diesel oil due to its lower polluting potential and the possibility of being used in diesel engines without the need for significant modifications [23, 24].

Biodiesel (alkyl esters) can be obtained by several reaction routes. The most conventional of them is the transesterification of triglycerides (oils) with short-chain alcohols (Eqs. (1)–(3)), such as methanol and ethanol, with homogeneous alkaline catalysts [25]. However, the raw material needed for this reaction must have reduced levels of acidity and moisture to avoid saponification reactions [26]. Due to the high costs of obtaining feedstocks that meet these specifications and competition with the food industry, studies aimed at the production of biodiesel from residual oils with a high content of fatty acids have been published [27–30].

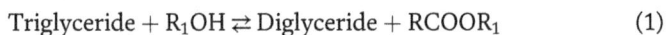

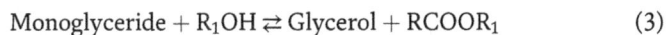

$$\text{Triglyceride} + R_1OH \rightleftarrows \text{Diglyceride} + RCOOR_1 \qquad (1)$$

$$\text{Diglyceride} + R_1OH \rightleftarrows \text{Monogliceryde} + RCOOR_1 \qquad (2)$$

$$\text{Monoglyceride} + R_1OH \rightleftarrows \text{Glycerol} + RCOOR_1 \qquad (3)$$

One of the alternatives to reduce the typical acidity of residual oils is to carry out a previous fatty acid esterification step [25] (Eq. (4)). However, in this process, with reaction and separation occurring in different units, chemical equilibrium limits the yield because the reaction that forms one of the products, water is reversible [31].

$$R_1COOH + R_2OH \rightleftarrows R_1COOR_2 + H_2O \qquad (4)$$

Therefore, the use of a reactive distillation column can be justified and may result in higher fatty acid conversions compared to those achieved in conventional systems. **Tables 1** and **2** provide an overview of the literature on esterification and transesterification reactions aimed at producing biodiesel.

In general, the main objective of the evaluated studies on esterification in reactive distillation columns is to increase the yield of the biodiesel production process by shifting the chemical equilibrium of the reaction. However, for transesterification studies, reactive distillation systems are considered mainly due to the possible reduction in the reaction time or in the purification costs of the biodiesel alkyl esters formed. This difference in purpose presumably originates from the necessity of milder transesterification reaction conditions when compared to the esterification route for the attainment of high biodiesel yields.

3. Mathematical Modeling

Mathematical modeling of a reactive distillation column was developed in [51] with considerations described in [52]. In their study, the authors assume the absence of chemical equilibrium in the stages and steady-state operation, with reaction rates being explicitly considered in the model of each stage, Murphree separation efficiency equal to 100% and feeding performed by single-phase streams.

Such methodology, presented in this section, was used in the studies of the references [2, 38, 39,53]. The nomenclature for the terms used in the equations is described in **Table 3**.

Ref.	Feedstock	Catalyst	Temperature	Simulation software	FFA conv.
[31][a]	Oleic acid	H$_2$SO$_4$ (1–3 FFA wt%)	130–150°C	—	79.6%
[32][a]	Dodecanoic acid	Solid acid catalyst (0–5 FFA wt%)	120–180°C	AspenOne 2004	~95.0%
[33][a]	Dodecanoic acid	Metal oxides (0–10 FFA wt%)	120–180°C	AspenOne 2004	~72.0%
[34][a]	Non-edible oil mixture	Tin (II) chloride (1–9 oil wt%)	40–60°C	Aspen Plus V8.8	78.3%
[35][b]	Dodecanoic acid	Sulfated Zirconia	130°C (FFA feed)	Aspen Plus V9	99.9%
[36][a]	Fatty acids mixture	Nb$_2$O$_5$ (5 FFA wt%)	90–170°C	Aspen Plus V7.3	96.0%
[37][b]	Fatty acids mixture	Sulfuric acid	417°C (FFA feed)	Aspen Plus	—
[38][b]	Fatty acids	Solid acid catalysts	58, 145, 207°C (FFA feed)	Fortran Algorithm	>98.0%
[39][b]	Hydrolyzed soybean oil	Niobium oxide	207°C (FFA feed)	Fortran Algorithm	>98.0%
[40][a]	Fatty acids	Sulfuric acid (0.33–0.66 wt%)	50–70°C	Aspen Plus V10	~90.0%

[a]*Experimental tests performed by the authors.*
[b]*No experimental tests performed by the authors.*

Table 1.
Reactive distillation studies aimed at the production of biodiesel through esterification.

Ref.	Feedstock	Catalyst	Temperature	Simulation software	Yield
[41][a]	Soybean oil	NaOH (0.5–1.5 oil wt%)	50°C (oil feed)	—	98.2%
[42][a]	Canola oil	KOH, KOCH$_3$ (0.73–1.83 oil wt%)	100–160°C (reboiler)	—	94.9%
[43][a]	Canola oil	KOH	95–150°C (reboiler)	—	94.4%
[44][a]	Cooking oil	12-Tungstophosphoric acid hydrate	20–30°C (oil feed)	—	93.8%
[45][a]	Palm oil	KOH (0.5–2 oil wt%)	85–120°C (reboiler)	—	92.3%
[46][b]	Triolein	NaOH	55°C (oil feed)	CHEMCAD/MATLAB	90.3%
[47][b]	Soybean oil	NaOH	25–90°C (oil feed)	HYSYS	~97.0%
[48][b]	Algal oil	H$_2$SO$_4$	<150°C (reboiler)	MATLAB	—
[49][b]	Trilinolein	NaOH/ Magnesium methoxide	60–150°C (column)	Aspen Plus	98.3%
[50][b]	Soybean oil	NaOH/CaO + Al$_2$O$_3$	60°C/25°C	Aspen Plus V8.4	—

[a]*Experimental tests performed by the authors.*
[b]*No experimental tests performed by the authors.*

Table 2.
Reactive distillation studies aimed at the production of biodiesel through transesterification.

Symbol	Description
$C_{i,j}$	Molar concentration
$C^L_{p,i}$	Molar specific heat in the liquid phase
E_j	Relationship between vapor and liquid streams
$F_{i,j}$	Feed molar flow
$f^{eq}_{i,j}$	Phase equilibrium function
$f^m_{i,j}$	Mass balance function
f^{vl}_j	Function that correlates liquid and vapor streams
$f^r_{k,j}$	Reaction rate function
H^I_j	Total enthalpy of the vapor stream
H^{II}_j	Total enthalpy of the liquid stream
$k_{k,j}$	Kinetic reaction constant
$n^I_{i,j}$	Molar flow of vapor
$n^{II}_{i,j}$	Molar flow of liquid
nc	Number of components
nr	Number of chemical reactions
$p^{sat}_{i,j}$	Liquid saturation pressure
P_j	Pressure
Q_j	Heat added or removed
R	Universal gas constant
R_j	Liquid withdrawal fraction
T_j	Temperature
T_{ref}	Reference temperature (298.15 K)

Symbol	Description
v^{II}_j	Liquid molar volume of pure compound
$x^I_{i,j}$	Molar fraction in vapor phase
$x^{II}_{i,j}$	Molar fraction in liquid phase
Z_j	Lateral vapor withdrawal fraction
$\alpha_{i,k}$	Kinetic order of reaction
Δh^{vap}_i	Molar enthalpy of vaporization
$\gamma^{vap}_{i,j}$	Activity coefficient in the liquid phase
$v_{i,k}$	Stoichiometric coefficient
$\xi_{k,j}$	Reaction rate

(references: i = component, j = column stage, k = reaction).

Table 3.
Nomenclature of terms used in mathematical modeling.

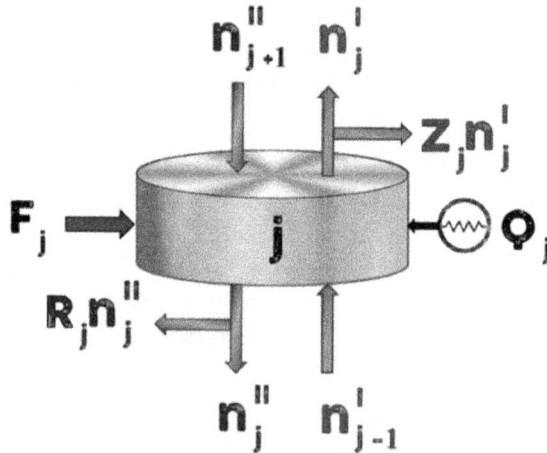

Figure 3.
Configuration of each stage j in the reactive distillation column. Source: [53].

The generic plate scheme adopted by the authors is represented in **Figure 3**. Eq. (5), which represents the mass balance of component i in stage j of the column as a residual function, is given by:

$$f^m_{i,j} = (R_j + 1)n^{II}_{i,j} + (Z_j + 1)n^I_{i,j} - \left(n^{II}_{i,j+1} + n^I_{i,j-1} + F_{i,j} + \sum_{k=1}^{nr} v_{i,j}\xi_{k,j} \right) = 0 \quad (5)$$

Assuming that the streams that leave the stage are in phase equilibrium, Eq. (6) relates the mole fractions in the liquid and vapor phases:

$$f^{eq}_{i,j} = \ln \left(x^I_{i,j}P_j \right) - \ln \left(x^{II}_{i,j}\gamma^{II}_{i,j}P^{sat}_{i,j} \right) = 0 \quad (6)$$

In this expression, the Poynting correction and the fugacity coefficient of the pure saturated compounds are neglected. In addition, and the vapor phase is considered to be an ideal gas mixture as a consequence of the assumption that the column operates at low pressure, close to atmospheric conditions.

The equation for the rate of reaction k in stage j is represented by Eq. (7), which can be expressed as a residual equation by applying the logarithm function (Eq. (8)).

$$\xi_{k,j} = k_{k,j} \prod_{i=1}^{nc} C_{i,j}^{\alpha_{i,k}} \tag{7}$$

$$f_{k,j}^r = \ln k_{k,j} + \sum_{i}^{nc} \alpha_{i,k} \ln \left(\frac{x_{i,j}^{II}}{v_j^{II}} \right) - \ln \xi_{k,j} = 0 \tag{8}$$

Assuming that the molar volume of the liquid phase is that of an ideal solution and describing Eq. (8) as a function of the activity coefficients of the components in the liquid phase, Eq. (9) is obtained.

$$f_{k,j}^r = \ln \left(k_{k,j} \right) + \sum_{i}^{nc} \alpha_{i,k} \ln \left(x_{i,j}^{II} + \gamma_{i,j}^{II} \right) - \ln \left(\xi_{k,j} \right) = 0 \tag{9}$$

Eq. (10), which describes the energy balance of stage j, is needed to calculate the temperature, which is different at each stage of the reactive distillation column. Positive and negative values of Q_j correspond to the heat being supplied to or removed from the column, respectively.

$$f_j^h = (R_j + 1)H_j^{II} + (Z_j + 1)H_j^I - \left(H_{j+1}^{II} + H_{j-1}^I + H_{F_j} + Q_j \right) = 0 \tag{10}$$

The ratio between the molar flows of vapor and liquid leaving each stage of the column is represented using Eq. (11). This equation is intended to make the condenser and reboiler specifications more flexible by associating the relationship between the liquid and vapor streams that leave the column stages.

$$E_j = \frac{(Z_j + 1)\sum_{i=1}^{nc} n_{i,j}^I}{(R_j + 1)\sum_{i=1}^{nc} n_{i,j}^{II}} \tag{11}$$

When written in the residual form, as in Eq. (12), the equation has the following form:

$$f_j^{vl} = (Z_j + 1) \sum_{i=1}^{nc} n_{i,j}^I - E_j(R_j + 1) \sum_{i=1}^{nc} n_{i,j}^{II} = 0 \tag{12}$$

The values of the E_j parameter for each form of operation of the condenser and reboiler (partial or total) are shown in **Table 4**.

In the study developed by [2, 36, 37, 48], all cases of simulated reactive distillation column configurations used a partial reboiler and total condenser ($E_1 \neq 0$ and $E_N = 0$).

Solving the set of equations that describe a reactive distillation column is an arduous task, and rigorous mathematical models aimed at a computer simulation of this type of equipment were not developed until the 1970s [54].

Reboiler (stage 1)		Condenser (stage N)	
Partial	**Total**	**Partial**	**Total**
$Z_1 = 0$	$Z_1 \neq 0$	$Z_N = 0$	$Z_N = 0$
$R_1 = 0$	$R_1 = 0$	$R_N = 0$	$R_N \neq 0$
$E_1 \neq 0$	$E_1 \to \infty$	$E_N \neq 0$	$E_N = 0$

Table 4.
Characteristics of column heaters (reboiler and condenser).

In recent decades, commercial software that has specific models and algorithms for reactive distillation operations has been widely used, as shown previously in **Tables 1** and **2**. The simulations developed in the subsequent section, referring to case studies applied to biodiesel production and co-product valuation, use the RADFRAC module present in the commercial Aspen Plus software, which solves the equations of mass balance, energy balance, phase equilibrium and the sum of molar fractions (MESH) [55] through the "inside-out" algorithm [54].

4. Fatty acid esterification simulation

4.1 Methodology

4.1.1 Kinetic parameters estimation

The kinetic parameters for the esterification of fatty acids (FFA) present in corn distillers oil from DDGS (dried distillers grains with solubles) were estimated by a model fitting of the FFA conversion data (**Table 5**) obtained by our group. The reaction (Eq. (13)) was carried out at the temperatures of 150, 175 and 200°C, with ethanol and $NbOPO_4$ (catalyst), following a molar alcohol:FFA ratio of 10:1 and catalyst load of 10% (FFA mass).

The methodology applied aims to estimate the pre-exponential factor (k_0) and the activation energy (E_a) of the reaction. To fit the kinetic parameters, the objective function to be minimized is the squared difference between the experimental values of the FFA conversion and those calculated with a reversible pseudo-homogeneous model (Eq. (14)). The reaction rate (r_F) equation was applied to Eq. (15), which describes a batch reactor in terms of the FFA conversion (x) in a given time (t).

A Nelder–Mead [56] simplex algorithm and a 4th order Runge–Kutta [57] method were used to perform the objective function minimization and conversion equation integration steps, respectively. The reaction rate (r_F) was obtained through the model and the specific reaction rate constants k (L/mol.s) were expressed by the Arrhenius equation (Eq. (16)). Through this methodology, the parameters E_a and k_0 for the reaction rate constants of the forward and reverse esterification reactions were estimated.

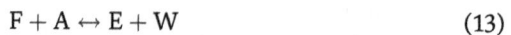

$$F + A \leftrightarrow E + W \tag{13}$$

$$-r_F = k_1 C_F C_A - k_2 C_E C_W \tag{14}$$

$$\frac{dx}{dt} = \frac{r_F}{C_{F(t=0)}} \tag{15}$$

$$k = k_0 e^{-\frac{E_a}{RT}} \tag{16}$$

Temperature (°C)	Time (min)	FFA conversion (%)
150	15	1.68
	30	3.99
	60	5.25
	120	9.68
	180	12.23
	240	19.44
	360	28.39
175	15	10.13
	30	15.36
	60	19.36
	120	24.21
	180	34.74
	240	43.06
	360	37.88
200	15	9.46
	30	17.97
	60	29.41
	120	38.35
	180	45.89
	240	48.53
	360	49.55

Table 5.
FFA conversion of the esterification reaction kinetic tests.

In these equations,
T = Absolute temperature at which the kinetic test was performed (K).
R = Universal gas constant (J/K.mol).
F = Fatty Acids (FFA).
A = Alcohol (ethanol).
E = Ethyl Esters (FAEE).
W = Water.
C_n = Concentration of compound n (mol/L).
t = time (s).

4.1.2 Process simulation

The compounds defined for the simulation of the fatty acid esterification with ethanol were specified in the Aspen Plus V.12 process simulator, with the fatty acid and oil fraction represented by oleic acid and triolein, respectively. A similar approach was used by other researchers as a simplification of the numerous components of the acid and oil fraction of the feedstock [58–60]. The NRTL thermodynamic model [61] was selected to evaluate the activity coefficients of the components of the reaction mixture and the NRTL binary interaction parameters missing from the simulator database were estimated directly through the Aspen Plus estimation tool that uses the UNIFAC model [62].

The flowsheet developed for the process simulation is shown in **Figure 4** and consists of two columns, the first being responsible for the reactive distillation of the reactants fed to the process (C-EST) and the second for removing approximately 95% of the ethyl esters (FAEE) produced (C-DIST).

The simulated reactive distillation column has 22 total stages, of which 14 compose the reactive zone (5 to 18), while the C-DIST column consists of 10 total stages. The columns operating parameters are presented in **Tables 6** and **7**. Both the distillation columns have kettle-type reboilers, however the C-EST column is equipped with a total condenser, while the C-DIST with a partial condenser to separate the ethyl esters from the remaining oil and excess ethanol. It is noteworthy that the liquid phase composition and temperature profile graphs, as well as the conversions obtained follow the data referring to the process after the optimization described later.

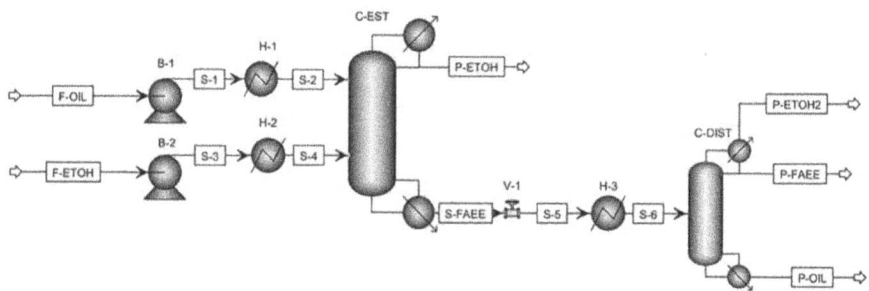

Figure 4.
Flowsheet of the FFA esterification process (F = feed, P = product, S = intermediate stream, H = heat exchanger, B = pump, V = valve, C = column).

Parameters	Before optimization
Stages	22
Oil feed stage	5
Ethanol feed stage	18
Absolute pressure (bar)	4
Distillate: feed molar ratio	0.62
Reflux molar ratio	0.08

Table 6.
Reactive distillation column operating parameters (C-EST).

Parameters	Distillation column
Stages	10
Feed stage	4
Absolute pressure (bar)	0.03
Distillate: feed molar ratio	0.51
Reflux molar ratio	1.5

Table 7.
Distillation column operating parameters (C-DIST).

Stream	F-OIL	F-ETOH
Temperature (°C)	20.00	20.00
Absolute pressure (bar)	1.00	1.00
Enthalpy (kW)	−4058.33	−2461.46
Mass Flow (kg/h)		
Oleic acid (FFA)	900	—
Ethanol	—	1467.86
Triolein	5100	—
Ethyl oleate (FAEE)	—	—
Water	—	—

Table 8.
Properties of the oil feed and ethanol streams (F-OIL and F-ETOH).

The H-1 and H-2 heat exchangers are responsible for heating the oil (F-OIL) and ethanol (F-ETOH) streams up to 200°C and 50°C, respectively, shown in **Table 8**, while the pumps B-1 and B-2 increase the pressure of the feed streams from 1 bar to 10 bar.

4.1.3 Optimization of the reactive distillation column

The optimization of the process parameters of the esterification reactive distillation column was performed using the MATLAB® R2020b software by implementing the MEIGO package (Metaheuristics for Bioinformatics Global Optimization) [63], an optimization supplement for global optimal search which can be used to optimize industrial processes [64, 65]. The results obtained through Aspen Plus simulations were provided to MATLAB, where the optimization algorithm was performed, and new obtained values of the variables evaluated were used to carry out new simulations iteratively.

The Particle Swarm Optimization (PSO) method was applied to minimize the objective function that describes the conversion of fatty acids (Eq. (17)), starting from an initial population of 50 particles (solution vectors) defined by the algorithm in the pre-defined search intervals.

For the simulation, the varied parameters were molar reflux ratio, internal pressure, molar ratio between distillate stream and total feed, and oil and ethanol feed stages. As restrictions, the reboiler temperature, the recovery of the desired product (ethyl esters) at the bottom of the column and the feed stages of the reagents were evaluated with Eqs. (18)–(20). The reboiler temperature upper limit was defined as 200°C to avoid degradation of the reagents or products and excessive use of the hot utility. It is observed that the minimization of the negative value of the conversion corresponds to the maximization of its positive value.

$$\min(conversion(\%)) = -\left(1 - \frac{Prod_{Bot}}{FFA_{in}}\right) \times 100 \tag{17}$$

$$T_{Reb} \leq 200°C \tag{18}$$

$$Rec_{Prod} = \frac{Prod_{Bot}}{Prod_{Top} + Prod_{Bot}} \geq 0.99 \tag{19}$$

$$FS_{Oil} \leq FS_{EtOH} \tag{20}$$

In these equations,

FFA_{in} = Molar flow of fatty acids in the feed stream (kmol/h).

T_{Reb} = Reboiler temperature (°C).

Rec_{Prod} = Desired product fraction recovered at the column bottom.

$Prod_{Bot}$ = Molar flow of the desired product at the column bottom (kmol/h).

$Prod_{Top}$ = Molar flow of the desired product at the column top (kmol/h).

FS_{Oil} = Acid oil (corn distillers oil) feed stage.

FS_{EtOH} = Ethanol feed stage.

4.2 Results

4.2.1 Kinetic parameters fitting

The kinetic constants obtained through the discussed methodology are presented in **Table 9**, with the direct reaction of ethyl esters formation indicated by the subscript "1" and the reverse reaction of fatty acids formation indicated by the subscript "2". **Figure 5** shows the comparison between the experimental and calculated conversions, along with the R^2 coefficient of the fit for each temperature.

Observing the results presented, it is noted that the data fitting at 200°C presented a high coefficient of determination, while the data fitting at 175°C obtained a reduced R^2. However, as the temperature in the reactive section of the esterification column is, on average, close to 195°C, it was concluded that due to the excellent results achieved in the data fitting at 200°C, the use of the estimated

Parameter	Value
$K_{0,1}$(L/mol.s)	252.786
$K_{0,2}$(L/mol.s)	207.093
Ea_1 (J/mol)	51,357.1
Ea_2 (J/mol)	39,244.1

Table 9.
Estimated kinetic constants for the esterification reaction.

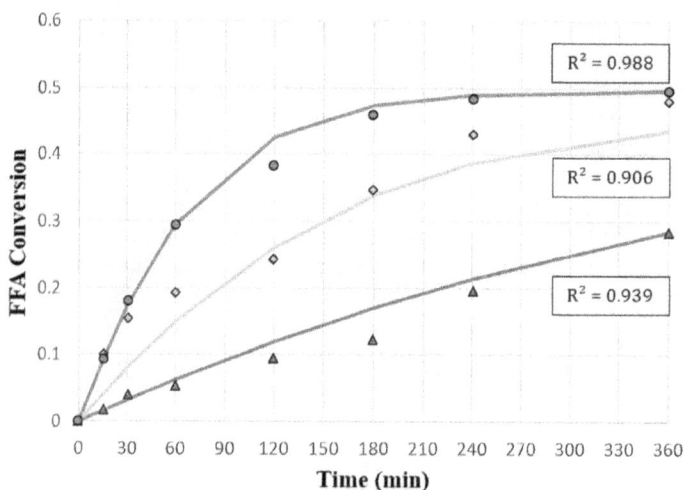

Figure 5.
Experimental (−) and calculated FFA conversion at 150°C (Δ), 175°C (◇) and 200°C (○).

kinetic parameters would not hinder the development of a simulation faithful to the real behavior of the reaction.

4.2.2 Esterification of fatty acids

The composition and temperature profiles along the stages of the reactive distillation column (column C-EST in **Figure 4**) are presented in **Figures 6** and 7.

The liquid phase composition profile of the C-EST column (**Figure 6**) indicates that the major component for all stages with values higher than 5 (closer to the bottom) is triolein, while in the others there is a predominance of oleic acid, ethyl oleate (FAEE), ethanol and water, since only negligible amounts of triolein are evaporated along the column, as seen in the vapor phase mass composition profile. Additionally, there is a significant increase in the fraction of ethanol and water in the liquid state in the first stage due to the use of a total condenser in the reactive distillation column.

The composition of the streams that characterize the main products of the process (S-FAEE, P-OIL and P-FAEE) are presented in **Table 10** and, based on the simulation results, there is a final fatty acids conversion (mol) of 83.97% inside the

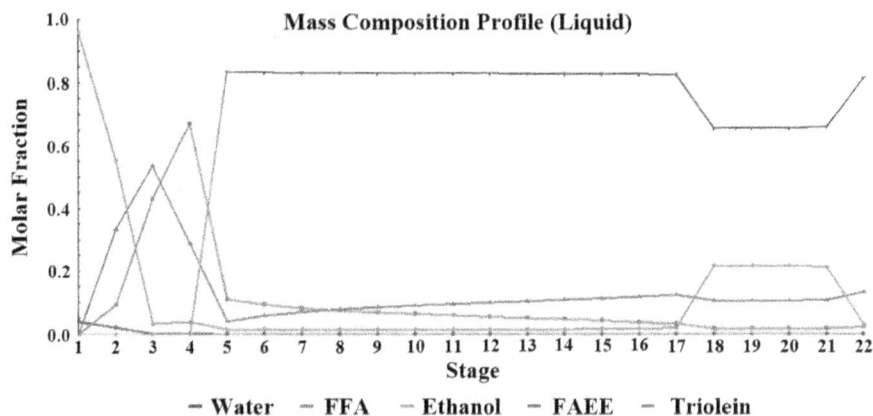

Figure 6.
Liquid phase mass composition profile (C-EST).

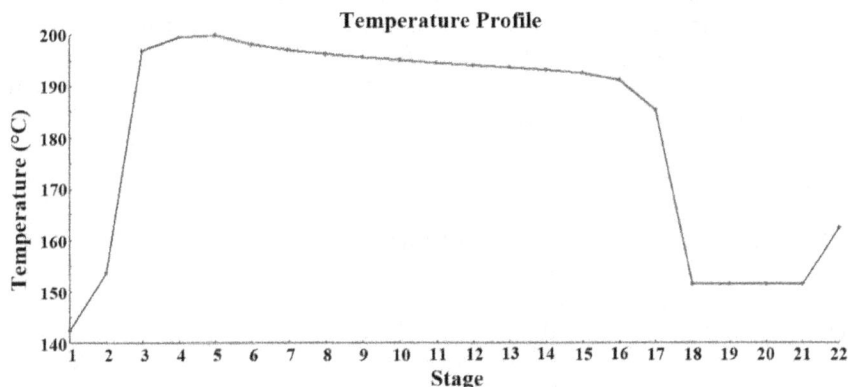

Figure 7.
Column temperature profile (C-EST).

Stream	S-FAEE	P-FAEE	P-OIL
Temperature (°C)	162.39	90.00	305.08
Absolute pressure (bar)	8.13	0.03	0.03
Enthalpy (kW)	−3813.85	−506.62	−2532.30
Mass Flow (kg/h)			
Oleic acid (FFA)	144.22	11.80	132.41
Ethanol	186.96	0.58	—
Triolein	5100.00	—	5100.00
Ethyl oleate (FAEE)	830.34	780.97	48.43
Water	0.58	—	—

Table 10.
Composition of the main product streams.

Equipment	Energy demand (kW)
H-1	619.71
H-2	31.60
C-EST (condenser)	−295.57
C-EST (reboiler)	330.86
C-DIST (condenser)	−581.94
C-DIST (reboiler)	1095.36

Table 11.
Energy demand of the process equipment.

column, 94.00% of which is recovered in the P-FAEE stream, while 5.83% is recovered in the P-OIL stream. The remaining 0.17% of FAEE is located at the P-ETOH2 stream. The resulting stream of the desired product (P-FAEE) has a purity (FAEE) greater than 98%, resulting in an ester content superior to the value described in Brazilian and European specifications [66, 67].

In **Table 10**, it is possible to observe that there are still traces of ethyl esters present in the oil stream. However, this amount corresponds to less than 1% of the total mass fraction of the stream. P-OIL, therefore, was considered to be non-significant. Furthermore, of the 900 kg/h of FFA fed to the process, only 132.41 kg/h remain, characterizing a reduction of 85.29% of the total fatty acid mass. Finally, the energy demands for H-1, H-2, condensers and reboilers of columns C-EST and C-DIST are presented in **Table 11**.

4.2.3 Optimization of the reactive distillation column

Table 12 shows the limits and initial estimates for the variables evaluated for the optimization of the esterification process. **Table 13** displays the constraints imposed on the reboiler temperature, ethyl ester recovery fraction (FAEE), and conversion. The values chosen as initial estimates were obtained by manually setting different values for the reflux molar ratio, condenser pressure, and distillate feed molar ratio, and adopting the best result obtained.

The results obtained are shown in **Figure 8**, with a maximum conversion of 83.97% and the final values of the variables are added to **Table 14**.

Variable	Molar reflux ratio	Condenser pressure (bar)	Distilled molar ratio: feed	Oil feed stage	Ethanol feed stage
Lower Limit	0.005	0.01	0.50	5	5
Upper Limit	2	10	0.70	18	18
Initial Estimate	0.08	4	0.62	5	18

Table 12.
Lower, upper limits and initial estimates for the variables evaluated in the esterification reaction optimization process.

Restrictions	Reboiler temperature (°C)	Recovery fraction of FAEE	Conversion (%)
Lower Limit	−273.15	0.99	0
Upper Limit	200	1.00	100

Table 13.
Initial constraints for the response variables for the variables evaluated in the esterification reaction optimization process.

An additional simulation performed in a CSTR reactor achieved an FFA conversion of 51.06%, while the maximum average conversion in the kinetic tests (200°C) was 49.55%. The simulated CSTR operated at a constant temperature of 200°C with the residence time of 3 h (same duration of the experimental tests) and was fed with streams following equal mass flows and compositions to the RDC column feed streams. Thus, the optimization results represent a significant improvement of 64.45% and 69.46% compared to the CSTR and experimental tests, respectively, inferring that the use of a reactive distillation column could be beneficial to the process.

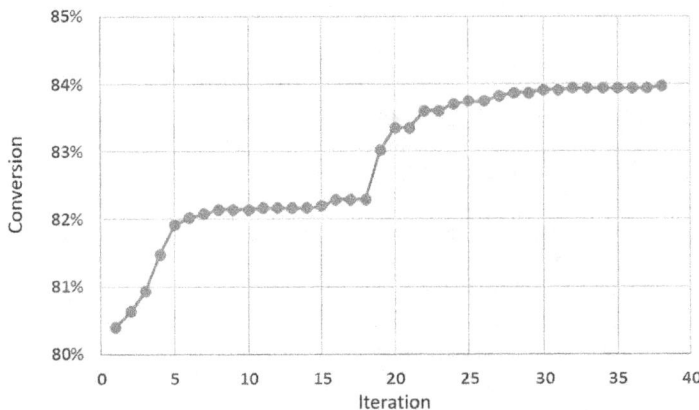

Figure 8.
Evolution of the FFA conversion as a function of the number of optimization iterations.

Variable	Molar reflux ratio	Condenser pressure (bar)	Distilled molar ratio: feed	Oil feed stage	Ethanol feed stage
Result	0.1130	8.1314	0.6806	5	18

Table 14.
Response vector of input variables for the esterification reaction optimization process.

5. Solketal production simulation

5.1 Justification

As biodiesel production increases so do the production of glycerol as for each liter of biodiesel produced, approximately 100 mL of crude glycerol are obtained [68]. Among the transformation processes for glycerol to viable chemical intermediates, glycerol ketalization for the production of solketal has gained prominence. Solketal can be used as an additive to increase the octane and fluid dynamic properties of the fuel. The addition of up to 5% by volume of solketal to gasoline leads to a significant decrease in gum formation [69]. With this motivation, this study aims to simulate the operation of a reactive distillation column for the production of solketal from glycerol with acetone using heterogeneous catalysis, with high conversion of reagents and separation of the components of the reaction.

5.2 Methodology

The applied methodology considers the ketalization reaction of glycerol (G) with acetone (A), forming solketal (S) and water (W). The reaction is considered reversible and elementary, being described by Eq. (21):

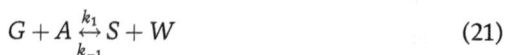

$$G + A \underset{k_{-1}}{\overset{k_1}{\rightleftharpoons}} S + W \tag{21}$$

A pseudo-homogeneous model was used to describe the reaction kinetics through a system of differential equations of concentration over time, at different temperatures, in which the kinetic constants of the direct and inverse reaction are represented, respectively, by k_1 and k_{-1} (L/mol.s), while the molar concentrations (mol/L) of the species involved are given by C_G, C_A, C_S and C_W (Eq. (22)).

$$-r_G = k_1 C_G C_A - k_{-1} C_S C_W \tag{22}$$

The solution of the system of differential equations using a 4th order Runge Kutta method [57] and the fitting of the kinetic parameters, k_1 and k_{-1}, and subsequent estimation of the Arrhenius equation parameters were performed by a Nelder–Mead simplex algorithm [56]. The experimental data used was retrieved from the study of [70].

The kinetic parameters evaluated were later used to predict the solketal formation reaction in a reactive distillation column, using the rigorous RADFRAC distillation model of the Aspen Plus commercial simulation software. The system considered in this study is shown in **Figure 9**.

Using the estimated kinetic parameters, the glycerol ketalization reaction for the production of solketal was modeled in the Aspen Plus software. For the process simulation, the pressure inside the column was set at 10 atm. The feeding of the 13-stage column, RDC in **Figure 9**, are streams GLI-02 and ACE-02, originated from the heating of the currents GLI-01 and ACE-01 up to 95°C and 55°C, by the heat exchangers H1 and H2, respectively.

The ACE-01 stream is composed only of acetone, while GLI-01 contains 80% glycerol and 20% water by mass, disregarding other components such as methanol or dissolved salts normally present in glycerol from biodiesel production processes [71]. The products of reactive distillation are characterized by TOP-P and BOT-P streams, which correspond, respectively, to the streams rich in the most volatile and least volatile substances in the process.

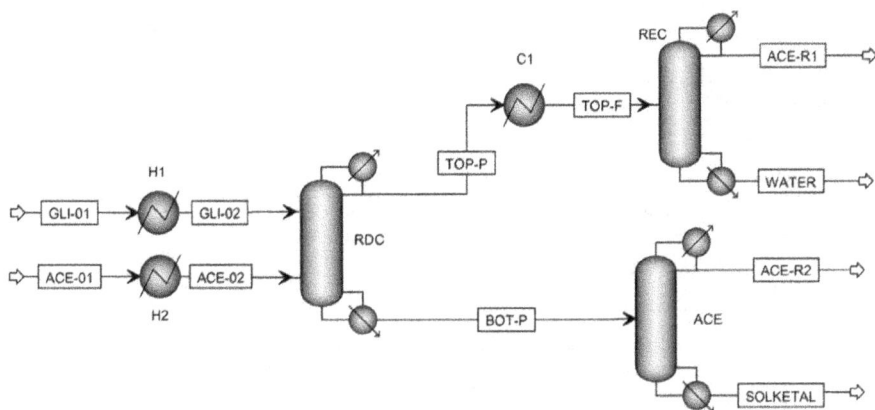

Figure 9.
Flowsheet of the solketal production process used in this study.

5.3 Results

Figure 10 shows the concentrations as a function of time according to the fitted kinetic parameters data.

Analyzing **Figure 10**, it is observed that the curves generated using the fitted parameters represented the experimental data satisfactorily. **Table 15** presents the process specifications obtained after a sensitivity analysis, aiming to simulate a column with optimal operating conditions. **Figure 11** shows the composition profile in the liquid phase as a function of the column stage number (1 = condenser and 13 = reboiler).

The conversion of glycerol obtained for the operational conditions defined for the simulation was 98.2%, indicating the reaction occurred inside the column.

The SOLKETAL stream in **Figure 9** has 99.53% solketal and the WATER stream consists of 99.82% water, on a mass basis. Thus, the simulations show that the

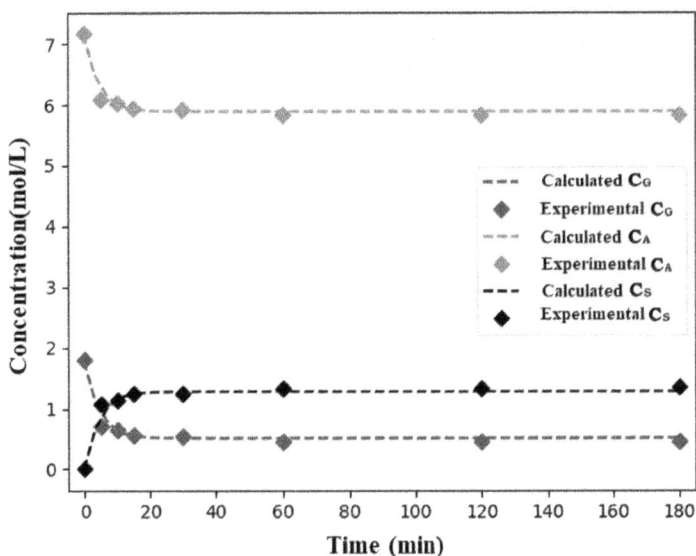

Figure 10.
Experimental and calculated concentrations (80°C).

Parameter	Description
Number of stages	13
Condenser type	Total
Reboiler type	Kettle
Molar reflux ratio	0.69
Reboiler/condenser heat duty	55.000 / -43.723 cal/s
Column pressure	10 bar
Glycerol feed	3rd stage
Acetone feed	11th stage
Feed properties	95 and 55°C - 1 bar
Glycerol feed molar flow	2.500 kmol/h
Water feed molar flow	0.625 kmol/h
Acetone feed molar flow	15.000 kmol/h
Ketalization reaction stages	3 to 11

Table 15.
RDC column specifications.

methodology employed results in a high purity solketal product stream with solketal conversion superior to 98%. However, additional studies are needed to assess the effect of possible intermediate reactions on the process yield.

6. Conclusions

In this chapter, a general introduction regarding reactive distillation technology and its application to the biodiesel production process was presented. A literature-based mathematical model to describe reactive distillation columns was discussed, along with experimental and simulation studies developed by the authors of this chapter, using commercial software such as Aspen Plus.

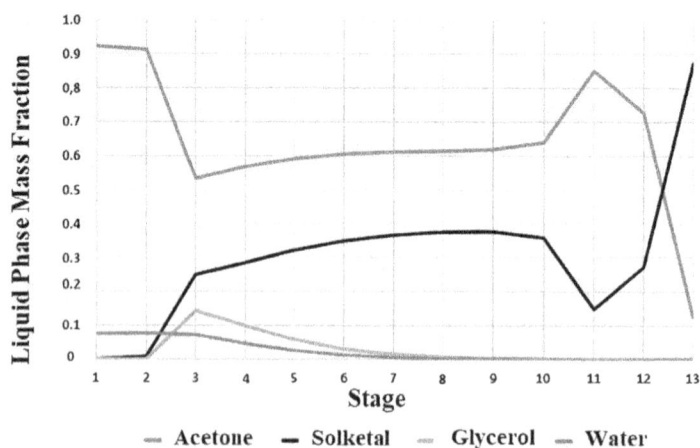

Figure 11.
RDC column liquid phase composition profile.

In the case study of biodiesel production through the esterification of a low-cost feedstock, the application of an optimized reactive distillation column promoted an improvement of approximately 70% about FFA conversion. The resulting product stream attained purity above 98% in relation to alkyl esters. Additionally, the production of solketal aiming at the valorization of a co-product of the biodiesel production process (glycerol), was studied through the development of a flowsheet in the Aspen Plus simulator, resulting in a solketal stream with purity above 99%.

The results obtained through the developed studies indicate that the reactive distillation technology, applied to fatty acid esterification reactions for the production of biodiesel and ketalization of glycerol for the production of solketal, is promising and attractive in technical terms, however, further studies are necessary to analyze the economic feasibility of both processes.

Acknowledgements

The authors thank UTFPR and Sinochem Petroleum Brazil Limited (project 001/2019) for financial support.

Author details

Guilherme Machado[1*], Marcelo Castier[2], Monique dos Santos[3], Fábio Nishiyama[1], Donato Aranda[4], Lúcio Cardozo-Filho[5], Vladimir Cabral[5] and Vilmar Steffen[6]

1 Federal Technological University of Paraná, Londrina, Brazil

2 German Paraguayan University, San Lorenzo, Paraguay

3 Sinochem Petroleum Brazil Limited, Rio de Janeiro, Brazil

4 Federal University of Rio de Janeiro, Rio de Janeiro, Brazil

5 State University of Maringá, Maringá, Brazil

6 Federal Technological University of Paraná, Francisco Beltrão, Brazil

*Address all correspondence to: guilhermed@utfpr.edu.br

IntechOpen

References

[1] Gotze L, Bailer O, Moritz P, von Scala C. Reactive distillation with KATAPAK®. Catalysis Today. 2001;**69**: 201-208. DOI: 10.1016/S0920-5861(01) 00370-4

[2] Machado GD. Simulação computacional da produção de biodiesel por hidroesterificação [thesis]. Maringá: Universidade Estadual de Maringá; 2011

[3] Taylor R, Krishna R. Modelling reactive distillation. Chemical Engineering Science. 2000;**55**:5183-5229. DOI: 10.1016/S0009-2509(00)00120-2

[4] Noeres C, Kenig EY, Górak A. Modelling of reactive separation processes: Reactive absorption and reactive distillation. Chemical Engineering and Processing. 2003;**42**: 157-178. DOI: 10.1016/S0255-2701(02) 00086-7

[5] Segovia-Hernández JG, Hernández S, Petriciolet AB. Reactive distillation: A review of optimal design using deterministic and stochastic techniques. Chemical Engineering and Processing. 2015;**97**:134-143. DOI: 10.1016/j. cep.2015.09.004

[6] Simasatikul L, Arpornwichanop A. Economic evaluation of biodiesel production from palm fatty acid distillate using a reactive distillation. Energy Procedia. 2017;**105**:237-243. DOI: 10.1016/j.egypro.2017.03.308

[7] Haydary J. Chemical Process Design and Simulation: Aspen Plus and Aspen Hysys Applications. 1st ed. Hoboken: John Wiley & Sons; 2019

[8] Backhaus AA. Apparatus for the manufacture of esters. US1400850 (Patent), 1921

[9] Backhaus AA. Apparatus for the production of esters. US1400851 (Patent), 1921

[10] Backhaus AA. Method for the production of esters. US1400852 (Patent), 1921

[11] Malone MF, Doherty MF. Reactive distillation. Industrial & Engineering Chemistry Research. 2000;**39**:3953-3957

[12] Agreda VH, Partin LR. Reactive distillation process for the production of methyl acetate. US4435595 (Patent), 1984

[13] Stankiewicz A, Moulijin J. Process intensification: Transforming chemical engineering. Industrial & Engineering Chemistry Research. 2002;**41**:1920-1924

[14] Sander S, Zuber L. Reactive distillation process and plant for obtaining acetic acid and alcohol from the hydrolysis of methyl acetate EP2502655A1 (Patent). 2012

[15] Panchal CB, Prindle JC. Method of producing high-concentration alkyl carbonates using carbon dioxide as feedstock. US9796656 (Patent). 2017

[16] Pacheco N, Dorato M, Jacquin M, Dastillung R, Couderc S. Method for producing butadiene from butanediols. WO2017102743A1 (Patent). 2017

[17] Gadewar SB, Vicente BC, Stoimenov PK. Production of butyl acetate from ethanol. US20160297737A1 (Patent). 2016

[18] Hoyme CA, Holocombe EF. Reactive distillation process for hydrolysis of esters. US6518465B2 (Patent). 2001

[19] Kronemayer H, Dahlhoff E, Lanver A. Process for preparing carboxylic esters by reactive distillation. WO2011131643A2 (Patent). 2011

[20] Almeida-Rivera CP. Designing Reactive Distillation Processes with

Improved Efficiency [Thesis]. Delft: Universidade Técnica de Delf; 2005

[21] Noeres C, Kening EY, Gorak A. Modelling of reactive separation processes: Reactive absorption and reactive distillation. Chemical Engineering and Processing. 2003;42: 157-178. DOI: 10.1016/S0255-2701(02) 00086-7

[22] Kularathne IW, Gunathilake CA, Rathneweera AC, Kalpage CS, Rajapakse S. The effect of use of biodiesel on environmental pollution – A review. International Journal of Renewable Energy Research. 2019;9:3

[23] Bhatia SK, Bhatia RK, Jeon J, Pugazhendhi A, Awasthi MK, Kumar D, et al. An overview on advancements in biobased transesterification methods for biodiesel production: Oil resources, extraction, biocatalysts and process intensification technologies. Fuel. 2021; 285:1-20. DOI: 10.1016/j. fuel.2020.119117

[24] Atadashi IM, Aroua MK, Aziz AA. Biodiesel separation and purification: A review. Renewable Energy. 2011;36: 437-443. DOI: 10.1016/ jrenene,2010.07.019

[25] Aranda DAG, Gonçalves JA, Peres JS, Ramos ALD, Melo CAR Jr, Antunes OAC, et al. The use of acids, niobium oxide, and zeolite catalysts for esterification reactions. Journal of Physical Organic Chemistry. 2009;22: 709-716. DOI: 10.1002/poc.1520

[26] Freedman B, Pryde EH, Mounts TL. Variables affecting the yields of fatty esters from transesterified vegetable oils. Journal of the American Oil Chemists' Society. 1984;61:1638-1643

[27] Hajjari M, Tabatabaei M, Aghbashlo M, Ghanavati H. A review on the prospects of sustainable biodiesel production: A global scenario with an emphasis on waste-oil biodiesel

utilization. Renewable and Sustainable Energy Reviews. 2017;72:445-464. DOI: 10.1016/j.rser.2017.01.034

[28] Gao X, Zhao R, Cong H, Na J, Shi Y, Li H, et al. Reactive distillation toward an ecoefficient process of biodiesel manufacture from waste oil: Pilot-scale experiments and process design. Industrial & Engineering Chemistry Research. 2020;59:14935-14946. DOI: 10.1021/acs.iecr.0c02343

[29] Haq IU, Akram A, Nawaz A, Zohu X, Abbas SZ, Xu Y, et al. Comparative analysis of various waste cooking oils for esterification and transesterification processes to produce biodiesel. Green Chemistry Letters and Reviews. 2021;14:461-472. DOI: 10.1080/17518253.2021.1941305

[30] Ding J, Xia Z, Lu J. Esterification and deacidification of a waste cooking oil (TAN 68.81 mg KOH/g) for biodiesel production. Energies. 2012;5:2683-2691. DOI: 10.3390/en5082683

[31] López-Ramírez MD, García-Ventura UM, Barroso-Munoz FM, Segovia-Hernández FO, Hernández S. Production of methyl oleate in reactive separation systems. Chemical Engineering and Technology. 2016;39: 271-275. DOI: 10.1002/ceat.201500423

[32] Kiss AA, Omota F, Dimian AC, Rothenberg G. The heterogeneous advantage: Biodiesel by catalytic reactive distillation. Topics in Catalysis. 2006;40:1-4. DOI: 10.1007/s11244-006-0116-4

[33] Kiss AA, Dimian AC, Rothenberg G. Biodiesel by catalytic reactive distillation powered by metal oxides. Energy & Fuels. 2008;22:598-604. DOI: 10.1021/ef700265y

[34] Kusumaningtyas RD, Prasetiawan H, Pratama BR, Prasetya D, Hisyam A. Esterification of non-edible oil mixture in reactive distillation

column over solid acid catalyst: Experimental and simulation study. Journal of Physical Science. 2018;**29**: 215-226. DOI: 10.21315/jps2018.29.s2.17

[35] Ali SS, Asif M, Basu A. Design and simulation of high purity biodiesel reactive distillation process. Polish Journal of Chemical Technology. 2019; **21**:1-7. DOI: 10.2478/pjct-2019-0022

[36] Banchero M, Gozzelino G. Nb2O5-catalyzed kinetics of fatty acids esterification for reactive distillation process simulation. Chemical Engineering Research and Design. 2015; **100**:292-301. DOI: 10.1016/j.cherd.2015. 05.043

[37] Cossio-Vargas E, Hernandez S. Segobia-Hernandez, Cano-Rodriguez MI. Simulation study of the production of biodiesel using feedstock mixtures of fatty acids in complex reactive distillation columns. Energy 2011;**36**: 6289-6297. 10.1016/j.energy.2011. 10.005

[38] Machado GD, Aranda DAG, Castier M, Cabral VF, Cardozo-Filho L. Computer simulation of fatty acids esterification in reactive distillation columns. Industrial & Engineering Chemistry Research 2011;**50**: 10176-10184. https:// dx.doi.org/ 10.1021/ie102327y

[39] Machado GD, Pessoa FLP, Castier M, Aranda DAG, Cabral VF, Cardozo-Filho L. Biodiesel production by esterification of hydrolyzed soybean oil with ethanol in reactive distillation columns: Simulation studies. Industrial & Engineering Chemistry Research 2013;**52**:9461-9469. https:// dx.doi.org/ 10.1021/ie400806q

[40] Margarida BR, Flores LI, Hamerski F, Voll FAP, Luz Jr LFL. Simulation, optimization and economic analysis of process to obtain esters from fatty acids. Biofuels, Bioproducts &

Biorefining. 2021;**15**:749-769. DOI: 10.1002/bbb.2186

[41] da Silva NL, Santander CMG, Batistella CB, Filho RM, Maciel MRW. Biodiesel production from integration between reaction and separation system: Reactive distillation process. Applied Biochemistry and Biotechnology. 2010; **161**:245-254. DOI: 10.1007/s12010-009-8882-7

[42] He BB, Singh AP, Thompson JC. Experimental optimization of a continuous-flow reactive distillation reactor for biodiesel production. Transactions of ASAE. 2005;**48**: 2237-2243. DOI: 10.13031/2013.20071

[43] He BB, Singh AP, Thompson JC. A novel continuous-flow reactor using reactive distillation for biodiesel production. Transactions of the ASABE. 2006;**49**:107-112. DOI: 10.13031/ 2013.20218

[44] Noshadi I, Amin NAS, Parnas RS. Continuous production of biodiesel from waste cooking oil in a reactive distillation column catalyzed by solid heteropolyacid: Optimization using response surface methodology (RSM). Fuel. 2012;**94**:156-164. DOI: 10.1016/j. fuel.2011.10.018

[45] Prasertsit K, Mueanmas C, Tongurai C. Transesterification of palm oil with methanol in a reactive distillation column. Chemical Engineering and Processing. 2013;**70**: 21-26. DOI: 10.1016/j.cep.2013.05.011

[46] Agarwal M, Singh K, Chaurasia SP. Simulation and sensitivity analysis for biodiesel production in a reactive distillation column. Polish Journal of Chemical Technology. 2012;**14**:59-65. DOI: 10.2478/v10026-012-0085-2

[47] Simasatikul L, Siricharnsakunchai P, Patcharavorachot Y, Assabumrungrat S, Arpornwichanop A. Reactive distillation for biodiesel production

from soybean oil. Korean Journal of Chemical Engineering. 2011;**28**: 649-655. DOI: 10.1007/s11814-010-0440-z

[48] Mondal B, Jana AK. Techno-economic feasibility of reactive distillation for biodiesel production from algal oil: Comparing with a conventional multiunit system. Industrial & Engineering Chemistry Research. 2019;**58**: 12028-12040. DOI: 10.1021/acs.iecr.9b00347

[49] Boon-Anuwat KW, Aiouache F, Assabumrungrat S. Process design of continuous biodiesel production by reactive distillation: Comparison between homogeneous and heterogeneous catalysts. Chemical Engineering and Processing. 2015;**92**: 33-44. DOI: 10.1016/j.cep.2015.03.025

[50] Poddar T, Jagannath A, Almansoori A. Biodiesel production using reactive distillation: A comparative simulation study. Energy Procedia. 2015;**75**:17-22. DOI: 10.1016/j.egypro.2015.07.129

[51] Alfradique MF, Castier M. Automatic generation of procedures for the simulation of reactive distillation using computer algebra. Computers & Chemical Engineering. 2005;**29**: 1875-1884. DOI: 10.1016/j.comp chemeng.2005.04.002

[52] Chen F, Huss RS, Malone MF, Doherty MF. Simulation of kinetic effects in reactive distillation. Computers & Chemical Engineering. 2000;**24**:2457-2472. DOI: 10.1016/S0098-1354(00)00609-8

[53] Machado GD, de Souza TL, Aranda DAG, Pessoa FLP, Castier M, Cabral VF, et al. Computer simulation of biodiesel production by hydro-esterification. Chemical Engineering and Processing. 2016;**103**:37-45. DOI: 10.1016/j.cep.2015.10.015

[54] Gautam P, Singh PP, Mishra A, Singh A, Das SN, Suresh S. Simulation of reactive distillation column. International Journal of ChemTech Research. 2013;**5**:1024-1029

[55] Popken T, Steinigeweg S, Gmehling J. Synthesis and hydrolysis of methyl acetate by reactive distillation using structured catalytic packings: Experiments and simulation. Industrial & Engineering Chemistry Research. 2001;**40**:1566-1574. DOI: 10.1021/ie0007419

[56] Nelder J, Mead R. A simplex method for function minimization. The Computer Journal. 1965;**7**:308-313. DOI: 10.1093/comjnl/7.4.308

[57] Dormand JR, Prince PJ. A family of embedded Runge-Kutta formulae. Journal of Computational and Applied Mathematics. 1980;**6**:19-26. DOI: 10.1016/0771-050X(80)90013-3

[58] Hussain Z, Kumar R. Esterification of free fatty acids: Experiments, kinetic modeling, simulation & optimization. International Journal of Green Energy. 2018;**15**:629-640. DOI: 10.1080/15435075.2018.1525736

[59] Margarida BR, Flores LI, Hamerski F, Voll FAP, Luz Jr LFL. Simulation, optimization, and economic analysis of process to obtain esters from fatty acids. Biofuels, Bioproducts and Biorefining. 2021;**15**:749-769. DOI: 10.1002/bbb.2186

[60] Anene RC, Giwa A. Modelling, simulation and sensitivity analysis of a fatty acid methyl ester reactive distillation process using Aspen plus. International Journal of Engineering Research in Africa. 2016;**27**:36-50. DOI: 10.4028/www.scientific.net/JERA.27.36

[61] Renon H, Prausnitz JM. Local compositions in thermodynamic excesso functions for liquid mixtures. AICHE

Journal. 1968;**14**:135-144. DOI: 10.1002/aic.690140124

[62] Fredenslund A, Jones RL, Prausnitz JM. Group-contribution estimation of activity coefficients in nonideal liquid mixtures. AICHE Journal. 1975;**21**:1086-1099. DOI: 10.1002/aic.690210607

[63] Egea JA, Henriques D, Cokelaer T, Villaverde AF, Macnamara A, Danciu DP, et al. MEIGO: An open-source software suite based on metaheuristics for global optimization in systems biology and bioinformatics. BMC Bioinformatics. 2014;**15**:1-9. DOI: 10.1186/1471-2105-15-136

[64] Bonfim-Rocha L, Gimenes ML, Farias SHB, Silva RO, Esteller LJ. Multi-objective design of a new sustainable scenario for bio-methanol production in Brazil. Journal of Cleaner Production. 2018;**187**:1043-1056. DOI: 10.1016/j.jclepro.2018.03.267

[65] Luna R, López F, Pérez-Correa JR. Minimizing methanol content in experimental Charentais alembic distillations. Jounal of Industrial and Engineering Chemistry. 2018;**57**:160-170. DOI: 10.1016/j.jiec.2017.08.018

[66] Ministério de Minas e Energia. Especificação do Biodiesel [Internet]. 2020. Available from: https://www.gov.br/anp/pt-br/assuntos/producao-e-fornecimento-de-biocombustiveis/biodiesel/especificacao-do-biodiesel [Accessed: 2021-10-04]

[67] Sajjadi B, Raman AA, Arandiyan H. A comprehensive review on properties of edible and non-edible vegetable oil-based biodiesel: Composition, specifications and prediction models. Renewable and Sustainable Energy Reviews. 2016;**63**:62-92. DOI: 10.1016/j.rser.2016.05.035

[68] Mota CJA, Pinto BP. Transformações catalíticas do glicerol para inovação na indústria química. Revista Virtual de Química. 2017;**9**:135-149. DOI: 10.21577/1984-6835.20170011

[69] Mota CJA, Rosenbach N Jr, Costa J, Silva F. Glycerin derivatives as fuel additives: The addition of glycerol/acetone ketal (solketal) in gasolines. Energy & Fuels. 2010;**24**:2733-2736. DOI: 10.1021/ef9015735

[70] Rossa V, Pessanha YSP, Diaz GC, Camara LDT, Pergher SBC, Aranda DAG. Reaction kinetic study of solketal production from glycerol ketalization with acetone. Industrial & Engineering Chemistry Research. 2016;**54**:479-488. DOI: 10.1021/acs.iecr.6b03581

[71] Mota CJA, Silva CXX, Gonçalves VLC. Gliceroquímica: novos produtos e processos a partir da glicerina de produção de biodiesel. Química Nova. 2009;**32**:639-648. DOI: 10.1590/S0100-40422009000300008

Chapter 8

Reactive Distillation Modeling Using Artificial Neural Networks

Francisco J. Sanchez-Ruiz

Abstract

The use of artificial intelligence techniques in the design of processes has generated a line of research of interest, in areas of chemical engineering and especially in the so-called separation processes, in this chapter the combination of artificial neural networks (ANN) is presented and fuzzy dynamic artificial neural networks (DFANN). Applied to the calculation of thermodynamic properties and the design of reactive distillation columns, the ANN and DFANN are mathematical models that resemble the behavior of the human brain, the proposed models do not require linearization of thermodynamic equations, models of mass and energy transfer, this provides an approximate and tight solution compared to robust reactive distillation column design models. Generally, the models must be trained according to a dimensionless model, for the design of a reactive column a dimensionless model is not required, it is observed that the use of robust models for the design and calculation of thermodynamic properties give results that provide better results than those calculated with a commercial simulator such as Aspen Plus (R), it is worth mentioning that in this chapter only the application of neural network models is shown, not all the simulation and implementation are presented, mainly because it is a specialized area where not only requires a chapter for its explanation, it is shown that with a neural network of 16 inputs, 2 hidden layers and 16 outputs, it generates a robust calculation system compared to robust thermodynamic models that contain the same commercial simulator, a characteristic of the network presented is the minimization of overlearning in which the network by its very nature is low. In addition, it is shown that it is a dynamic model that presents adjustment as a function of time with an approximation of 96–98% of adjustment for commercial simulator models such as Aspen Plus (R), the DFANN is a viable alternative for implementation in processes of separation, but one of the disadvantages of the implementation of these techniques is the experience of the programmer both in the area of artificial intelligence and in separation processes.

Keywords: reactive distillation, neural networks, dynamic fuzzy neural network, thermodynamics properties, design column, azeotropic mix

1. Introduction

Reactive distillation is a separation process that is implemented for the separation of complex mixtures because it combines a chemical reaction in a single piece of equipment, that is, one or more of the stages of the separation column has the function of a chemical reactor, in which the catalyzed or uncatalyzed reaction will

IntechOpen

be carried out, this type of process is implemented for mixtures that present azeo-tropes, these with very close boiling points that can be complex or require an excess of energy for the separation of the components, on some occasions the process is implemented for the purification of substances through a thermally integrated process, the reactive distillation process is carried out by mass transfer both in the liquid phase and in the vapor phase, or on the surface of the catalyst [1].

The calculation and design of a reactive distillation system introduce a term in the mass balances of the stages which makes it become reactive stages $M_{i,j}$ Eq. (1) thus becomes.

$$M_{i,j} = n_{L,j-1}x_{i,j-1} + n_{v,j+1}y_{i,j+1} + n_{Fj}x_{F,i,j}$$

$$(n_{lj} + n_{SL,j})x_{i,j} - (n_{vj} + n_{SVj})y_{i,j} - (V_{L,H})_j \sum_{n=1}^{n_{Rx}} v_{i,n}r_{j,n} = 0 \tag{1}$$

where $(V_{LH})_j$ is the volumetric liquid holdup at stage j, $v_{i,n}$ is the stoichiometric coefficient of component i in reaction n, $r_{j,n}$ rate of reaction n on stage j, and n_{Rx} is the number of chemical reactions.

The modification of stage energy balance is in the definition of Q_j in Eq. (2), where the heat of reaction is included.

$$H_j = n_{L,j-1}h_{Lj-1} + n_{vj+1}h_{vj+1} + n_{F,j}h_{F,j} -$$

$$(n_{L,j} + n_{SL,j})h_{Lj} - (n_{vj} + n_{SVj})h_{vj} - Q_j = 0 \tag{2}$$

In these equations, n represents mole flow, x mole fraction in the liquid phase, y mole fraction in the vapor phase, K equilibrium constant, h molar enthalpy, and Q heat flow. The subscript i represents a component, j stage, L liquid, V vapor, SV side vapor, SL side liquid, F feed, and N last stage, respectively [1].

A first approximation is carried out using a mathematical model based on equations in steady-state, these equations are taken as the basis for modeling in a dynamic state, which is necessary for the implementation of fuzzy dynamic artificial neural networks, therefore; the models presented in this chapter are those that are implemented for artificial intelligence systems.

Dynamic fuzzy neural networks have been implemented to solve non-linear mathematical models. In the areas of process engineering, it has been implemented in temperature control systems. In this chapter the use of artificial intelligence techniques to calculate a temperature is shown. Reactive column, where azeotropes are present in a ternary mixture.

2. Artificial neural networks

Artificial neural networks arise from the analogy that is made between the human brain and computer processing, from the first analyzes of the human brain carried out by Ramón y Cajal [2]. This analogy is made from the aspects of the neural structure to processing capacity.

Artificial neural networks are mathematical models that attempt to mimic the capabilities and characteristics of their biological counterparts. Neural networks are made up of simple calculation elements, all of them interconnected with a certain topology or structure, such as neurons called perceptron's, which are the simplest elements of a network. The basic model of a neuron is formed as observed by the following elements (**Figure 1**) [3, 4]:

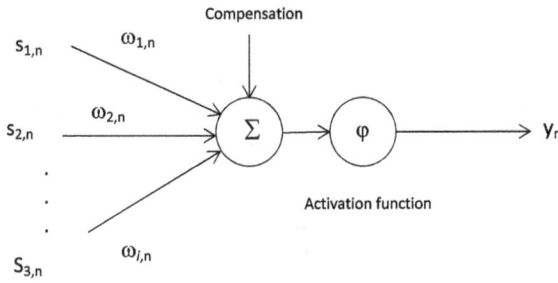

Figure 1.
Elementary neuron.

- A set of synapses, which are the inputs of the neuron weighted by a weight.

- An added that simulates the body of the neuron and obtains the level of excitation.

- The activation function generates the output if the excitation level is reached and restricts the output level, thus avoiding network saturation.

- The output of the neuron is given by the expression:

$$y_i = \varphi \left(\sum_{j=1}^{n} w_{ij} s_j + w_{i0} \right) \tag{3}$$

where n indicates the number of inputs to neuron i and φ denote the excitation function [5, 6]. The argument of the activation function is the linear combination of the inputs of the neuron. If we consider the set of inputs and the weights of neuron i as a vector of dimension (n + 1), the expression is concluded as follows:

$$y_i = \varphi \left[W_i^T s \right] \tag{4}$$

where

$$s = [-1, s_1, s_2, \dots, s_n]^T \tag{5}$$

$$w_i = [w_{i0}, w_{i1}, \dots, w_{in}]^T \tag{6}$$

Neural networks are classified into static and dynamic networks, the first of these have a broader field of application mainly due to their characteristic of no change for time, dynamic networks are applied more specifically to problems that present changes for time [7, 8].

The static and dynamic neural networks have the characteristics of similar this in mathematical structures, training in addition to principles of architectures of the same neural networks, the most commonly used networks are the so-called multi-layer neural networks this mainly because they resemble structures of the human brain, they can be networks with forwarding propagation but also networks with backward propagation. The selection of the same depends on the type of study system and the application of the network [9, 10]. For prediction systems of break-down curves in adsorption processes, the so-called multilayer neural networks with forwarding propagation are generally used, this is because it is not necessary to use a backward propagation of information as a means of comparison, the latter are most commonly applied in control processes [11–13].

2.1 Multilayer networks

A multilayer network has a defined structure, it consists of an input layer, hidden layers and an output layer (**Figure 2**), the definition of structure of a multilayer neural network has the characteristic of avoiding problems with the training of the network which generally results in prediction problems of the breakdown curve of the adsorption process, the process of establishing the architecture of the neural network is mainly based on a series of trial and error although in some of the cases if the programmer is an expert this is significantly reduced, mainly to the fact that there are already established mechanisms to determine architecture, Hecht-Nielsen (1989) [14] based on Kolmogorov's theorem [15–17], "The number of neurons in the hidden layer does not need to be greater than twice the number of inputs" using this theorem, the neuron approximation equation is established in the hidden layer [18, 19] Eq. (7).

$$h = \left(\frac{2}{3}\right)(n + m) \qquad (7)$$

where h represents the number of neurons in the hidden layer, n number of inputs and m is defined as the number of hidden layers, using this rule a stop parameter is established which means that the number of neurons in the hidden layer will never be required. More than twice the number of entries $h < 2n$. When it comes to a multilayer network with a single hidden layer, it is recommended that the number of neurons is 2/3 of the number of inputs [20, 21].

The next step in structuring a neural network is the establishment of the excitation functions, these functions can propagate the information and use them for the training of the same network, the information introduced into the network is known as synaptic weights, alluding to the synapses of biological neurons [22, 23]. The excitation functions are of different types, their choice depends on the type of process to be modeled, each excitation function is found in each of the neurons, both in the hidden layers and in the inputs and outputs. The most commonly used functions are the type function: tangential sigmoidal Eq. (8), logarithmic Eq. (9) and radial base type functions Eq. (10), this last function is one of the complex ones generally used for systems dynamic, in non-dynamic processes it can be used but this increases the computing time and information processing mainly because it becomes more specific in its application Eqs. (11)–(16) [24–27].

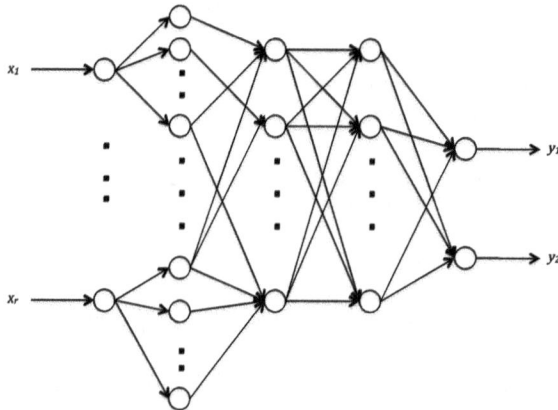

Figure 2.
Multilayer network.

$$\varphi = \frac{e^{-w_i} + e^{w_i}}{e^{-w_i} - e^{w_i}} \tag{8}$$

$$\varphi = \frac{1}{1 + e^{-w_i}} \tag{9}$$

$$\varphi = \sum_{i=1}^{N} w_i \Phi(\|w - w_{ci}\|) \tag{10}$$

Gaussian function

$$\Phi(w) = e^{w_i^2} \tag{11}$$

Multi-quadratic function

$$\Phi(w) = \sqrt{1 + w_i^2} \tag{12}$$

Reciprocad multi-quadratic function

$$\Phi(w) = \frac{1}{\sqrt{1 + w_i^2}} \tag{13}$$

Armonic-poli function

$$\Phi(w) = w_i^k \quad k = 1, 3, 5, \ldots \tag{14}$$

$$\Phi(w) = w_i^k \ln(w_i) \quad k = 2, 4, 6, \ldots \tag{15}$$

Slim quadratic function

$$\Phi(w) = w_i^2 \tag{16}$$

Once the excitation function or also called the transfer function has been selected, the neural network is trained for which there are different types of training, as with the selection of the architecture, the training is also selected by trial and error but if the experienced programmer can initiate selection with training for a certain type of neural structure, the most commonly used training is backward propagation (BP) training [28, 29], other types of training most used are Levenberg-Maquart (LM) and Broyden Fletcher Goldfarb Shanno (BFGS). Backward propagation training is the basis for all other training, for that reason, only this type of training will be discussed [30, 31].

The error signal at the output of neuron j in iteration k is defined by:

$$e_j(k) = d_j(k) - y_j(k) \tag{17}$$

The instantaneous value of the error is defined for neuron j, the sum of the instantaneous errors squared is formulated as:

$$\varepsilon(n) = \frac{1}{2} \sum_{j \in h_{out}}^{l} e_j^2(k) \tag{18}$$

where h_{out} is the set of output neurons, $h_{out} = \{1, 2, \ldots, l\}$. The average error (e_{av}) it is obtained by averaging the instantaneous errors corresponding to the N training pairs.

$$\varepsilon_{av}(n) = \frac{1}{N}\sum_{k=1}^{N}\varepsilon(k) \tag{19}$$

The objective is to minimize ε_{av} with respect to weights. You need to calculate $\Delta w_{ji}(k)$.

$$\frac{\partial\varepsilon(k)}{\partial w_{ji}(k)} \tag{20}$$

$$\frac{\partial\varepsilon(k)}{\partial w_{ji}(k)} = \frac{\partial\varepsilon(k)}{\partial e_j(k)}\frac{\partial e_j(k)}{\partial y_j(k)}\frac{\partial y_j(k)}{\partial v_j(k)}\frac{\partial v_j(k)}{\partial w_{ji}(k)} \tag{21}$$

$$v_j(k) = \sum_{i=0}^{p}w_{ji}(k)y_j(k) \tag{22}$$

$$y_j(k) = \varphi(v_j(k)) \tag{23}$$

The components to calculate the error are defined as follows.

$$\frac{\partial\varepsilon(k)}{\partial e_j(k)} = e_j(n) \tag{24}$$

$$\frac{\partial e_j(k)}{\partial y_j(k)} = -1 \tag{25}$$

$$\frac{\partial y_j(k)}{\partial v_j(k)} = \varphi_j(v_j(k)) \tag{26}$$

$$\frac{\partial v_j(k)}{\partial w_{ji}(k)} = y_j(k) \tag{27}$$

The gradient of the error is determined with Eq. (35).

$$\frac{\partial\varepsilon(k)}{\partial w_{ji}(k)} = -e_j(k)\varphi_j(v_j(k))y_j(k) \tag{28}$$

2.2 Dynamic fuzzy artificial neural network (DFANN)

The DFANNs use an excitation function based on asymmetric radial type function, which implies that the system behaves like a Takegi-Sugeon model (T-S) which has a characteristic pulse of a radial function bias. For the inputs of the fuzzy neural network, it is necessary to establish the limits of the inputs within a known interval, when this type of network is applied in the determination of properties, the inputs must be defined within known ranges, to avoid overlearning of the same artificial neural network. The structure of DFNN is shown in **Figure 3**, which is similar to the traditional models of artificial neural networks with the difference of the propagation of the synaptic weights in the radial basis excitation function, which can be biased or unbiased, the structure is defined below [32]:

Layer 1: Each node represents an input linguistic variable.

Layer 2: Each node represents a membership function (MF) which is in the form of Gaussian function Eq. (29).

$$MF_{ij} = \exp\left[-\frac{(x_i - c_{ij})^2}{\sigma_j^2}\right] \quad i = 1, \dots, r \text{ and } j = 1, \dots u \quad (29)$$

Where MF_{ij} is the jth membership function of x_i, c_{ij} is the center of jth Gaussian membership function of x_i and σ_j is the width of the jth Gaussian membership function of x_i, r is the number of input variables and u is the number of membership function [32].

Layer 3: Each node represents a possible IF, part for fuzzy rules. For the jth rule R_j, its output is:

$$OR_j = \exp\left[-\frac{\sum_{j=1}^{r}(x_i - c_{ij})^2}{\sigma_j^2}\right] \quad j = 1, \dots, u \quad (30)$$

$$OR_j = \exp\left[-\frac{\|X - C_j\|^2}{\sigma_j^2}\right] \quad (31)$$

Where $X = (x1, \dots, xr)$ and Cj is the center of the jth Radial Basic Function (RBF) unit.

Layer 4: Nodes as N (Normalized) nodes. The number of N nodes is equal to that of layer 3 the output of Nj is:

$$ON_j = \frac{OR_j}{\sum_{k=1}^{u} OR_k} = \frac{\exp\left[-\frac{\|X-C_j\|^2}{\sigma_j^2}\right]}{\sum_{k=1}^{u}\exp\left[-\frac{\|X-C_k\|^2}{\sigma_k^2}\right]} \quad (32)$$

Layer 5: Each node in this layer represents an output variable, which is the weighted sum of the incoming signals. Have:

$$y(x) = \sum_{k=1}^{u} ON_k w_{2k} = \frac{\sum_{k=1}^{u} w_{2k}\exp\left[-\frac{\|X-C_k\|^2}{\sigma_k^2}\right]}{\sum_{k=1}^{u}\exp\left[-\frac{\|X-C_k\|^2}{\sigma_k^2}\right]} \quad (33)$$

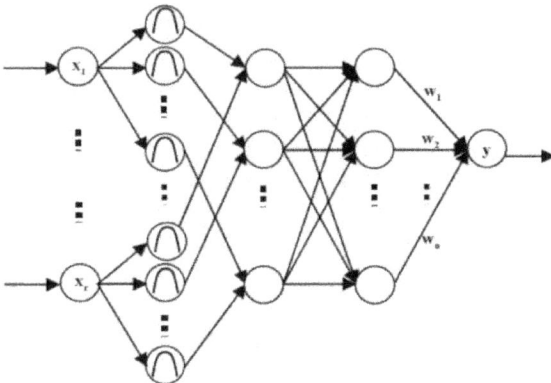

Figure 3.
Dynamic fuzzy artificial neural network (DFANN).

Where y is the value of an output variable and w_{2k} is the connection weight of each rule:

For the TSK (Takagi Sugeon and Kang).

$$w_{2k} = k_{j0} + k_{j1}x_1 + \dots + k_{jr}x_r \quad j = 1, 2, \dots, u \tag{34}$$

3. Methodology

3.1 Reactive distillation using neural networks artificial

Reactive distillation is implemented to separate mixtures of components that generally have more than one azeotrope, artificial human networks can be implemented to determine thermodynamic properties, for the solution of the differential equations of mass and energy transfer, the case study that presents artificial neural networks were implemented to determine the thermodynamic properties, as well as the solution for the mass transfer equations and the output compositions at the top of the column and the bottom (**Figure 4**).

3.2 Case study

A multicomponent mixture of ethanol, water, ethyl acetate, acetic acid and butanol is studied, the latter in a small proportion, it is observed that according to the multicomponent mixture, 2 azeotropic mixtures and azeotrope are formed in a ternary mixture of water-ethyl acetate-ethanol, which means that it is a complex reaction system, for the determination of thermodynamic properties as well as the design of a conventional distillation column.

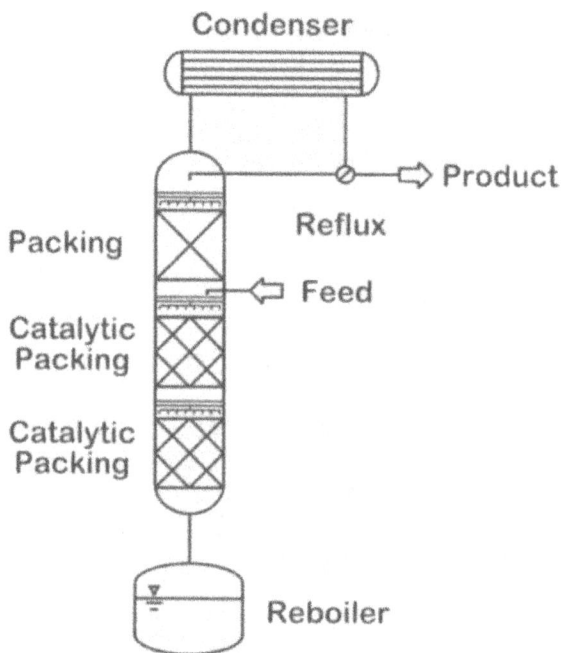

Figure 4.
Schematic distillation reactive [33].

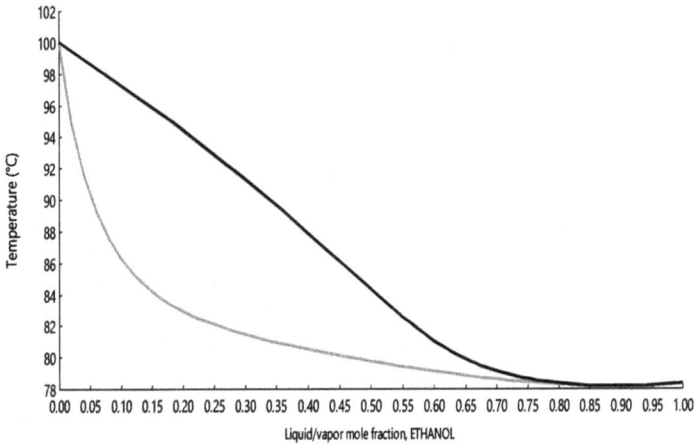

Figure 5.
Azeotrope ethanol-water.

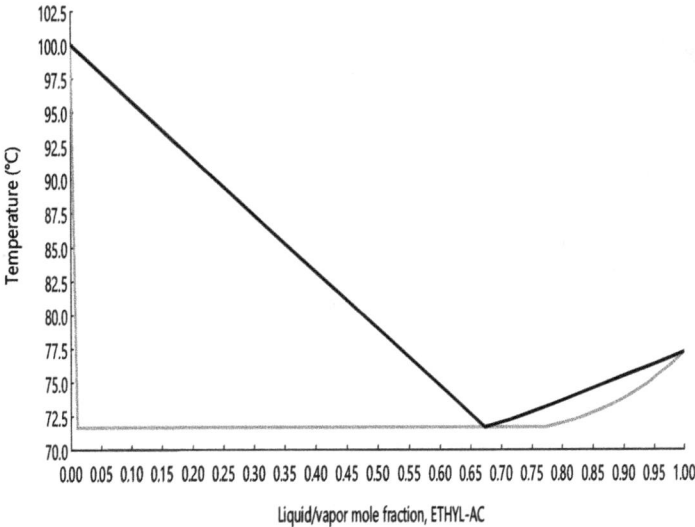

Figure 6.
Azeotrope ethyl acetate-water.

Figure 5 shows the binary azeotrope between ethanol-water, **Figure 6** shows the azeotrope between ethyl acetate-water, **Figure 7** between ethanol-ethyl acetate, at different temperatures; this implies that the conventional distillation column must be large in a number of plates, as geometry. **Figure 8** shows the ternary diagram of azeotrope formation.

In the compartment model (CM), one of the compartments is defined to consist of multiple single stages. Without loss of generality, balance equations for one single stage, the so-called sensitivity stage, can be replaced by the overall compartment balances. Assuming that stages Nc, 1 to Nc, 2 form compartment c (**Figure 9**), we obtain [1].

$$M_c = \sum_{i=Nc,1}^{Nc,i} M_i \tag{35}$$

$$x_c^j = \frac{1}{M_c} \sum_{i=Nc,1}^{Nc,i} M_i x_i^j \qquad j = Component \tag{36}$$

$$h_c^L = \frac{1}{M_c} \sum_{i=Nc,1}^{Nc,i} M_i h_c^L \tag{37}$$

Assuming the compartments to be sufficiently large, single-stage dynamics can be neglected compared to overall compartment dynamics. Consequently, single-stage balance equations are assumed stationary. Thus, the entire equation system for compartment c consists for stages $j = Nc, 1 ... Nc, 2$ and the steady-state versions for stages $j = Nc, 1 ... Nc, 2$ except for the sensitivity stage [1].

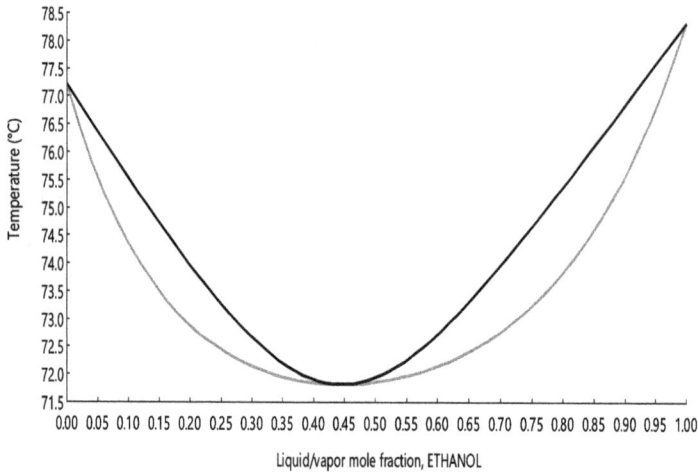

Figure 7.
Azeotrope ethanol-ethyl acetate.

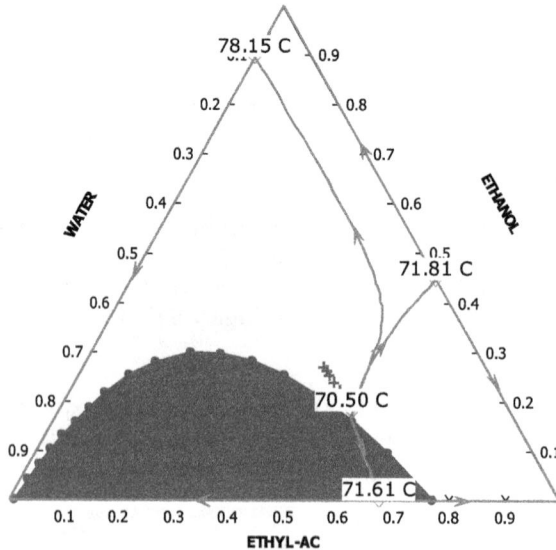

Figure 8.
Ternary diagram ethanol-ethyl acetate-water.

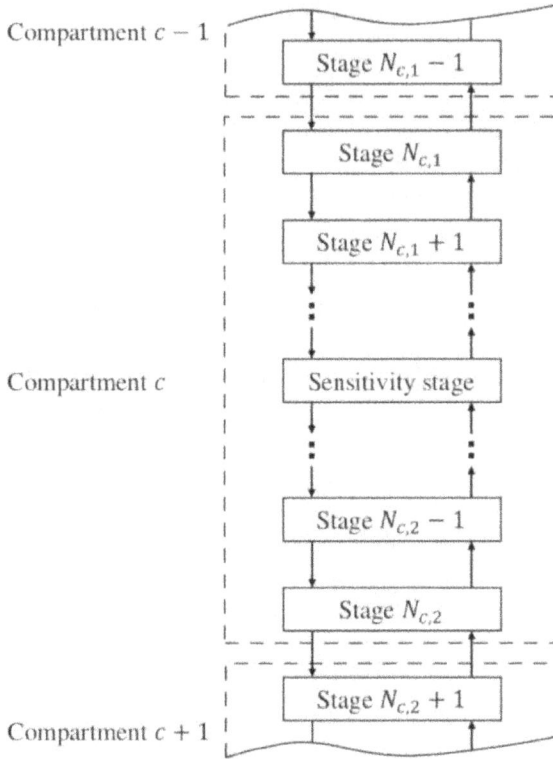

Figure 9.
Compartment model (CM). Dashed lines depict the compartment boundaries [33].

We derive the proposed model from the representation of the original compartment model as an:

$$\dot{x}(t) = \hat{f}(\hat{x}(t), \hat{u}(t), \hat{y}(t)) \qquad (38)$$

$$0 = \hat{g}(\hat{x}(t), \hat{u}(t), \hat{y}(t), \hat{z}(t)) \qquad (39)$$

In Eqs. (38) and (39), we introduce the following notation: differential compartment states are denoted by $x(t)$. Compartment inputs, which are handed over by the neighboring compartments, are denoted by $\hat{u}(t)$ which corresponds to states of column stages Nc, 1–1 *and* Nc, $i + 1$. Compartment outputs, which are required in the equation system of neighboring compartments, are denoted $y(t)$, which corresponds to the states of column stages Nc, 1 *and* Nc, i. The remaining (algebraic) compartment states are denoted $z(t)$.

When solving Eqs. (40) and (41), the main computational effort is spent in the solution of the highly nonlinear algebraic that mainly originates from the thermodynamic relations (Eqs. (38) and (39)). To reduce the computational effort, Linhart and Skogestad [34], propose interpolation between tabulated solved solutions.

$$\begin{pmatrix} \hat{y}(t) \\ \hat{z}(t) \end{pmatrix} = \hat{g}^{-1}(\hat{x}(t), \hat{u}) \qquad (40)$$

Sophisticated computer codes are readily available for efficient training of the ANNs. In particular, ANNs can also be fitted very efficiently to large data sets, which arise from a sampling of the high-dimensional input space.

$$\hat{x}(t) = \hat{f}(\hat{x}(t), \hat{u}(t), \hat{y}(t)) \tag{41}$$

$$\hat{y}(t) = \hat{g}_{ANN}(\hat{x}(t), \hat{u}(t)) \tag{42}$$

We highlight that model formulation Eqs. (41) and (42) is only one possibility of an ANN-based compartmentalization approach. The choice of this system, however, seems appealing as it shows an analogy to the dynamic modeling of simple flash units, that is, the model outputs can be calculated as an explicit function of the model inputs (Tx-flash or single-stage). Other possibilities for using ANNs exist as well. For instance, using a surrogate model for the ordinary differential equation (ODE) form thermodynamic system. Such are, the ANN could also be used to replace specific parts of mapping o searching thermodynamics properties.

Where a combination of ANN and CM result in a new model of reactive distillation, but new mathematical model show relationship between stochiometric and mass transfer, such relation to be.

$$T_i(t) = \sum_{i=Nc,1}^{Nc,2} OM_{c,i}T_{2i} = \frac{\sum_{k=1}^{m}\sum_{i=Nc,1}^{Nc,i} M_i T_{2,k} \exp\left[-\frac{T-T_i^2}{\varphi_k^2}\right] \exp\left[-\frac{M-M_i^2}{\varphi_k^2}\right]}{\sum_{k=1}^{m}\sum_{i=Nc,1}^{Nc,i} \exp\left[-\frac{T-T_i^2}{\varphi_k^2}\right] \exp\left[-\frac{M-M_i^2}{\varphi_k^2}\right]} \tag{43}$$

$$P_i(t) = \sum_{i=Nc,1}^{Nc,2} OK_{c,i}P_{2i} = \frac{\sum_{k=1}^{m}\sum_{i=Nc,1}^{Nc,i} K_i P_{2,k} \exp\left[-\frac{P-P_i^2}{\varphi_k^2}\right] \exp\left[-\frac{K-K_i^2}{\varphi_k^2}\right]}{\sum_{k=1}^{m}\sum_{i=Nc,1}^{Nc,i} \exp\left[-\frac{P-P_i^2}{\varphi_k^2}\right] \exp\left[-\frac{K-K_i^2}{\varphi_k^2}\right]} \tag{44}$$

Subsequently, for reactive distillation, an approach to reaction kinetics of study mixture is necessary, this is established by the following Eq. (45).

$$r(t) = k_1 \bar{x}_2 \bar{x}_0 - k_1 \bar{x}_1 \bar{x}_3 \tag{45}$$

where x_o represents the liquid fraction of acetic acid, x_1 of water, x_2 ethanol fraction, and x_3 ethyl acetate fraction.

Starting from the reaction kinetic equation, balances are established for each of the components as a function of time, for each stage of the separation process, that is, for each plate in the reactive column with DFANN.

$$M_j \frac{d\bar{x}_{i,j}}{dt} = +V\bar{y}_{i,j-1} - L\bar{x}_{i,j} - V\bar{y}_{i,j} + \zeta M_j R \tag{46}$$

$$\sum_{i,j}^{n} w_{ij} \prod_{ij}^{m} \bar{x}_{i,j}(t) M_j = \sum_{i}^{n}\sum_{j}^{m}\left[L\bar{x}_{i,j+1} + V\bar{y}_{i,j-1} - L\bar{x}_{i,j} - V\bar{y}_{i,j} + \zeta M_j R\right] \tag{47}$$

where i = 0, 2, 3 components and j = 1, ..., n number plates, $y_{i,j}$ is fraction steam in column ζ stoichiometric coefficient.

In condenser with DFANN

$$M_j \frac{d\bar{x}_{i,n}}{dt} = V\bar{y}_{i,n-1} - L\bar{x}_{i,n} - D_h\bar{x}_{i,n} + \zeta M_n R \tag{48}$$

$$\sum_{i,j}^{n} w_{ij} \prod_{ij}^{m} \bar{x}_{i,n}(t) M_j = \sum_{i}^{n}\sum_{j}^{m}\left[V\bar{y}_{i,n-1} - L\bar{x}_{i,n} - D_h\bar{x}_{i,n} + \zeta M_n R\right] \tag{49}$$

Reboiler with DFANN

$$\frac{d(M_0, \bar{x}_{i,0})}{dt} = L\bar{x}_{i,0} - V\bar{y}_{i,0} + \zeta M_0 R \tag{50}$$

$$\sum_{i,j}^{n} w_{i,j} \prod_{ij}^{m} \bar{x}_{i,0}(t) M_0 = \sum_{i}^{n} \sum_{j}^{m} \left[L\bar{x}_{i,0} - V\bar{y}_{i,0} + \zeta M_0 R \right] \tag{51}$$

where $\frac{dM_0}{dt} = L - V$ and $\bar{x}D_{c,j} = 1 - \sum_{i=Ac}^{Cc} \bar{x}_{i,j}$ from j = 0, 1, ... , n; the mole fraction of vapor meets the constraint.

$$\sum_{i=Ac}^{Ac} \bar{y}_{i,j} = \sum_{i=Ac}^{Pc} \frac{P_{i,j}\left(T_i \bar{x}_{i,j}\right)}{P_j} = 1. \tag{52}$$

where $D_{c,j}$ represents the amount of bottom distillate, likewise the partial pressures of the vapor and liquid phase are determined and M_j represents. **Table 1** shows the constants for the simulation.

The density with DFANN of the mixture is calculated by Eq. (53).

$$\rho_l = \sum_{i=1}^{m} \left[\prod_{d=1}^{n} (A_d B_d)^{(1-Tr)^{2/7}} \cdot w_{di,dj} \right]_i \tag{53}$$

The case study simulation was performed through modular programming in Aspen Plus®, using code linking in Matlab ®, the simulation parameters in Aspen Plus are shown in **Figures 10–16**, shows in a summarized way the parameters entered in a RadFrac column of Aspen, the calculation of properties was carried out with the binding of Matlab® and Aspen using both the NRTL methods for Aspen, as well as the DFANN methods in Matlab®.

3.3 Training dynamic fuzzy artificial neural network

The neural network has a structure of 16 inputs, two hidden layers with 12 neurons in each of the layers, and 16 output neurons, the training is based on an unsupervised training of the Quasi-Newton-Function (QSF) type, without backward propagation.

M_0 (kg/s)	4.798	k_1	2900exp(−7150/T(K))
M_j (kg/s)	0.0125	k_2	7380exp(−7150/T(K))

Table 1.
Constants for simulation.

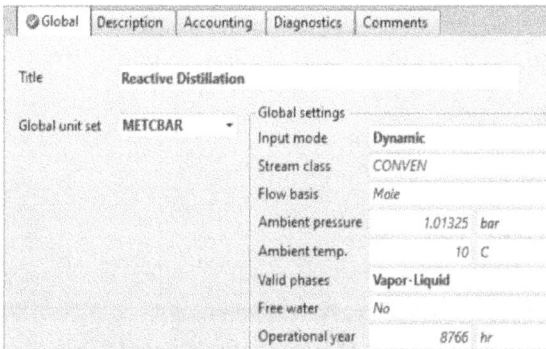

Figure 10.
Global design.

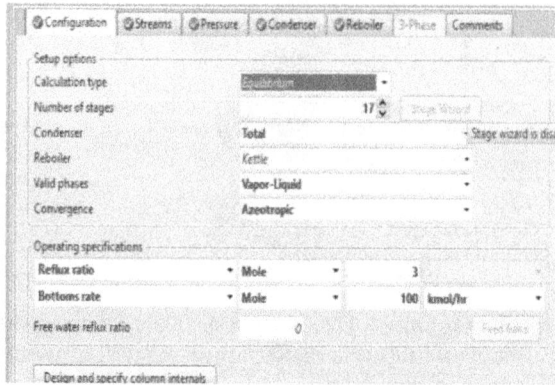

Figure 11.
Configuration.

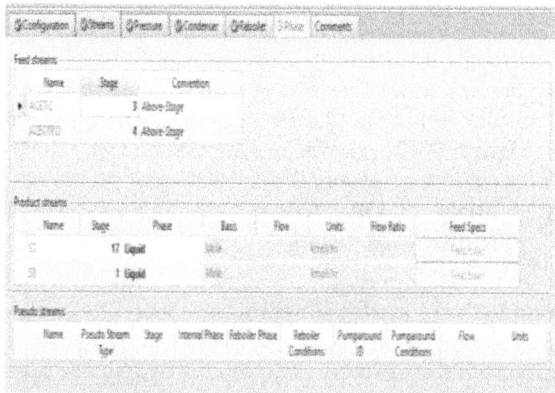

Figure 12.
Stages of feed.

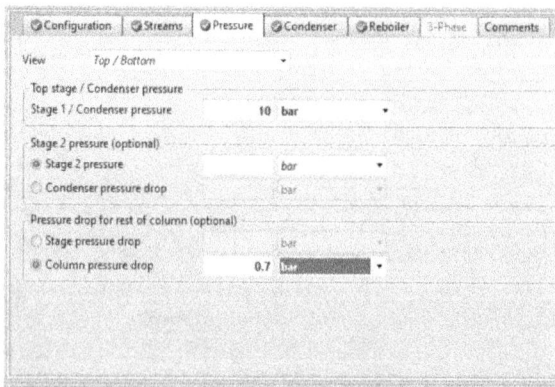

Figure 13.
Pressure design.

It is worth mentioning that the structure of the neural network was optimized using a supervised algorithm, which is based on the heuristic rule of 2n, where n is the number of inputs to the neural network, which implies that the minimum

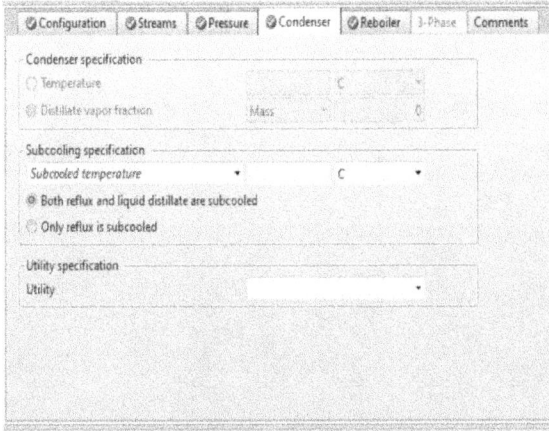

Figure 14.
Condenser design.

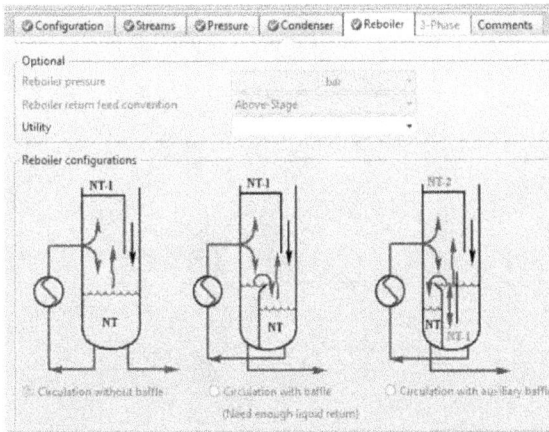

Figure 15.
Reboiler design.

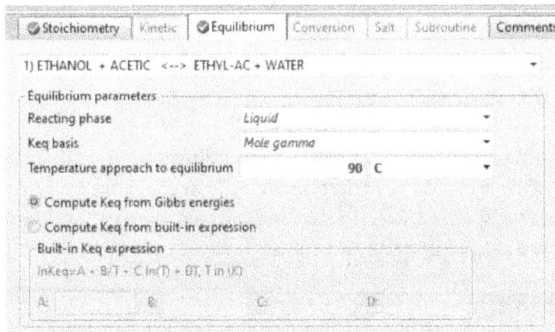

Figure 16.
Reaction equilibrium.

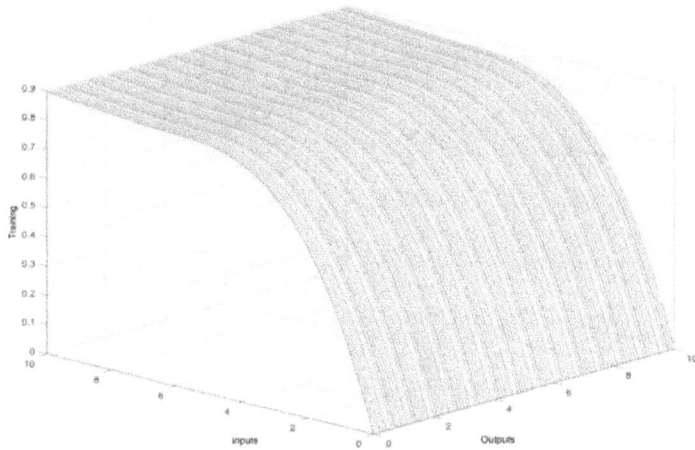

Figure 17.
Training surface.

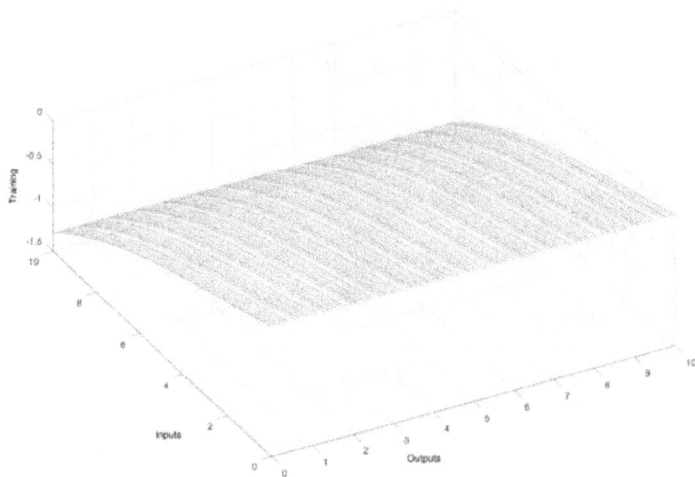

Figure 18.
Training response.

number of neurons present is sought. To avoid overlearning, **Figures 17** and **18** show the learning settings.

3.4 Results and discussion case study

Figure 19 schematically shows a distillation column using the commercial Aspen Plus® simulator, two streams are introduced, the acetic acid stream separated from the azeotropic mixture, this to facilitate the transfer of mass and energy, each stream is fed in one stage superior and in under stage, to facilitate the mentioned phenomena.

Simulations are performed using the same configuration with 17 separation stages (**Table 2**), the feeds were carried out in stages 4 and 5 respectively, the compositions of the dome and the bottom of the column are compared to determine the purity of the components, where y_i is the output composition in mole fraction, it

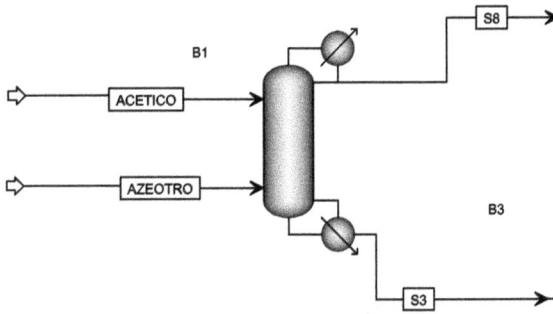

Figure 19.
Schematic of reactive distillation.

	Dome	Bottom	Dome	Bottom	Dome	Bottom
Component	yi (Aspen)	yi (Aspen)	yi(ANN)	yi(ANN)	yi(DFANN)	yi(DFANN)
Ethanol	0.9698	0.0302	0.9854	0.0146	0.9978	0.0022
Water	0.9519	0.0481	0.9723	0.0277	0.9822	0.0178
Acetic-Acid	0.1692	0.8308	0.2056	0.7944	0.3328	0.6672
Ethyl Acetate	0.9288	0.0712	0.9523	0.0477	0.9855	0.0145

Table 2.
Composition reactive distillation.

is observed that the column that simulates the process with artificial intelligence provides better results based on its ability to displace the azeotropes present.

The reaction that takes place in the reactive stages is as follows:

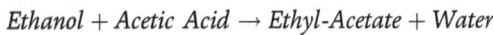

$$Ethanol + Acetic\ Acid \rightarrow Ethyl\text{-}Acetate + Water$$

In **Figure 20** the behavior of the liquid in each of the stages of the column is shown (**Figure 21**), it is observed that in the initial stages there is a transfer of both mass and energy, this can be verified in **Figure 22** where shows the temperature profile across the reactive column, in **Figures 23** and **24** represent the liquid and vapor profiles in each of the stages.

In the same simulations, comparisons were made between thermodynamic properties such as: Enthalpy and Entropy, these used in the thermodynamic models used.

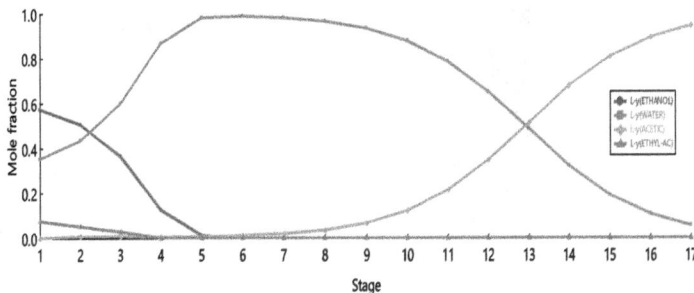

Figure 20.
Composition profile per stage using DFANN.

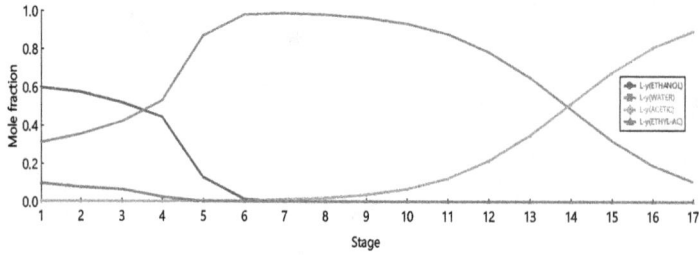

Figure 21.
Composition profile per stage using Aspen plus®.

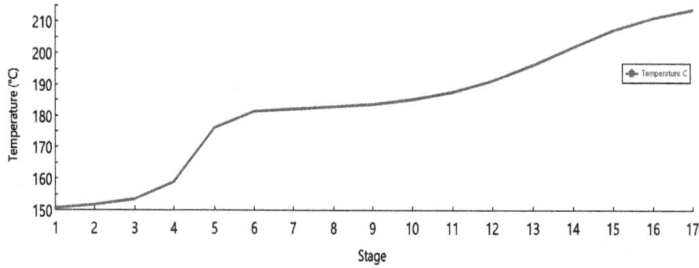

Figure 22.
Temperature profile per stage using DFANN.

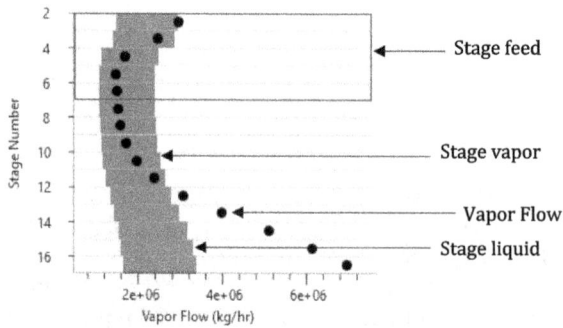

Figure 23.
Profile per stage vapor flow using DFANN.

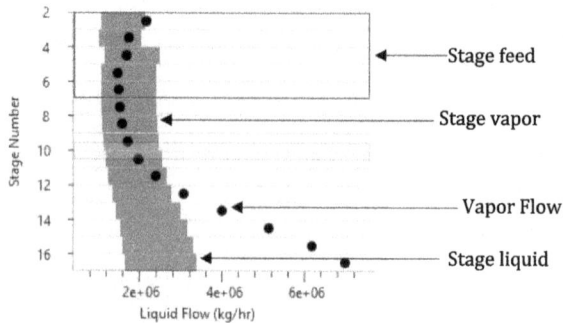

Figure 24.
Profile per stage liquid flow using DFANN.

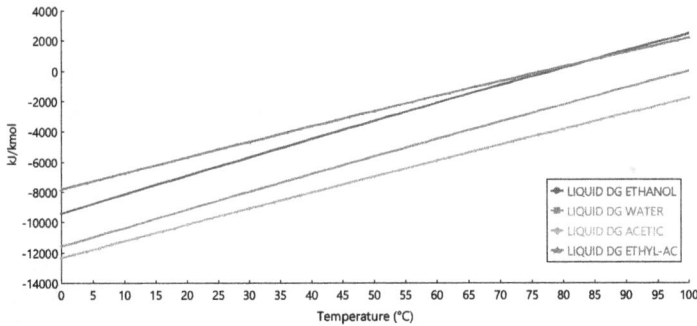

Figure 25.
Profile enthalpy DFANN.

In **Figure 25**, properties thermodynamic calculated using the intelligence algorithm are shown.

4. Conclusion

The implementation of artificial intelligence techniques such as artificial neural networks to the separation processes, provides promising results in matters of feasibility and process dynamics, in computational time the use of robust models for the calculation of properties compared to the use of networks. Similarly, with a significant decrease for neural network models, prediction-based fit and azeotrope separation based on variables such as temperature and pressure, neural networks provide better results compared to robust thermodynamic models with Aspen Plus®, which are models that in some cases implement statistical molecular mechanics. Fuzzy artificial neural networks adjust to the dynamics of the reactive column process, where separation of 99% is obtained, which implies that the azeotrope moves, in comparison with traditional models, adjusting the parameters according to the change in stoichiometry, one of the advantages in the ability to predict the change of azeotrope as a function of temperature and pressure; system, as well as the ability to establish the variables in permissible limits and limitations of the number of stages, without being a large design as the robust models mentioned, can give.

Author details

Francisco J. Sanchez-Ruiz
Environment Faculty, UPAEP University, Puebla, Mexico

*Address all correspondence to: franciscojavier.sanchez@upaep.mx

IntechOpen

References

[1] Haydary J. Chemical Process Design and Simulation: Aspen Plus and Aspen Hysys Applications. John Wiley & Sons; 2019

[2] Ramon Y, Cajal S. Textura del Sistema Nervioso del Hombre y de los Vertebrados. Vol. 2. Madrid: Nicolas Moya; 1904

[3] Yegnanarayana B. Artificial Neural Networks. PHI Learning Pvt Ltd; 2009

[4] Hopfield JJ. Artificial neural networks. Circuits and Devices Magazine, IEEE. 1988;4(5):3-10. DOI: 10.1109/101.8118

[5] Drew PJ, Monson JR. Artificial neural networks. Surgery. 2000;127(1):3-11. DOI: 10.1067/msy.2000.102173

[6] Abraham A. Artificial neural networks. In: Handbook of Measuring System Design. 2005

[7] Mäkisara K, Simula O, Kangas J, Kohonen T. Artificial Neural Networks. Vol. 2. Elsevier; 2014

[8] Narendra KS, Parthasarathy K. Identification and control of dynamical systems using neural networks. IEEE Transactions on Neural Networks. 1990;1(1):4-27. DOI: 10.1109/72.572089

[9] Gupta M, Jin L, Homma N. Static and Dynamic Neural Networks: From Fundamentals to Advanced Theory. John Wiley & Sons; 2004. DOI: 10.1002/0471427950

[10] Chiang YM, Chang LC, Chang FJ. Comparison of static-feedforward and dynamic-feedback neural networks for rainfall–runoff modeling. Journal of Hydrology. 2004;290(3):297-311. DOI: 10.1016/j.jhydrol.2003.12.033

[11] Pearlmutter BA. Learning state space trajectories in recurrent neural networks. Neural Computation. 1989; 1(2):263-269. DOI: 10.1162/neco.1989.1.2.263

[12] Basheer IA, Hajmeer M. Artificial neural networks: Fundamentals, computing, design, and application. Journal of Microbiological Methods. 2000;43(1):3-31. DOI: 10.1016/S0167-7012(00)00201-3

[13] Miller WT, Werbos PJ, Sutton RS. Neural Networks for Control. MIT Press; 1995. Available from: https://dl.acm.org/doi/abs/10.5555/104204

[14] Hecht-Nielsen R. Theory of the backpropagation neural network. In: Neural Networks. 1989, June, 1989. IJCNN., International Joint Conference on IEEE. pp. 593-605. DOI: 10.1016/B978-0-12-741252-8.50010-8

[15] Hecht-Nielsen R. Neurocomputing: picking the human brain. IEEE Spectrum. 1988;25(3):36-41. DOI: 10.1109/6.4520

[16] Hecht-Nielsen R. On the algebraic structure of feed forward network weight spaces. Advanced Neural Computers. 1990:129-135. DOI: 10.1016/B978-0-444-88400-8.50019-4

[17] Kůrková V. Kolmogorov's theorem and multilayer neural networks. Neural Networks. 1992;5(3):501-506. DOI: 10.1016/0893-6080(92)90012-8

[18] Hornik K, Stinchcombe M, White H. Multilayer feedforward networks are universal approximators. Neural Networks. 1989;2(5):359-366. DOI: 10.1016/0893-6080(89)90020-8

[19] Kolmogorov AN. The representation of continuous functions of many variables by superposition of continuous functions of one variable and addition.

Doklady Akademii Nauk SSSR. 1957;
114(5):953-956. Available from: http://
www.mathnet.ru/php/archive.phtml?
wshow=paper&jrnid=dan&paperid=
22050&option_lang=eng

[20] Yager RR, Kacprzyk, J. (Eds.). The
Ordered Weighted Averaging
Operators: Theory and Applications.
Springer Science & Business Media;
2012. Available from: https://link.
springer.com/chapter/10.1007/978-94-
009-0125-4_44

[21] Chen AM, Hecht-Nielsen R. On the
geometry of feedforward neural
network weight spaces. In: Artificial
Neural Networks, 1991, Second
International Conference on IET. 1991.
pp. 1-4

[22] Kohonen T. Self-organized
formation of topologically correct
feature maps. Biological Cybernetics.
1982;**43**(1):59-69

[23] Mehrotra K, Mohan CK, Ranka S.
Elements of Artificial Neural Networks.
MIT Press; 1997

[24] Rojas R. Neural Networks: A
Systematic Introduction. Springer
Science & Business Media; 2013

[25] Cybenko G. Approximation by
superpositions of a sigmoidal function.
Mathematics of Control, Signals and
Systems. 1989;**2**(4):303-314.
DOI: 10.1007/BF02134016

[26] Hashem S. Sensitivity analysis for
feedforward artificial neural networks
with differentiable activation functions.
In: Neural Networks, 1992 IJCNN,
International Joint Conference on IEEE.
Vol. 1. 1992. pp. 419-424. DOI: 10.1109/
IJCNN.1992.287175

[27] Shen W, Guo X, Wu C, Wu D.
Forecasting stock indices using radial
basis function neural networks
optimized by artificial fish swarm
algorithm. Knowledge-Based Systems.

2011;**24**(3):378-385. DOI: 10.1016/j.
knosys.2010.11.001

[28] White H. Artificial neural networks:
Approximation and learning theory.
Blackwell Publishers Inc; 1992

[29] Haykin SS, Haykin SS, Haykin SS,
Haykin SS. Neural Networks and
Learning Machines. Vol. 3. Upper
Saddle River: Pearson Education; 2009

[30] Leonard J, Kramer MA.
Improvement of the backpropagation
algorithm for training neural networks.
Computers & Chemical Engineering.
1990;**14**(3):337-341. DOI: 10.1016/
0098-1354(90)87070-6

[31] Goh ATC. Back-propagation neural
networks for modeling complex
systems. Artificial Intelligence in
Engineering. 1995;**9**(3):143-151.
DOI: 10.1016/0954-1810(94)00011-S

[32] Wu S, Er MJ, Liao J. A novel
learning algorithm for dymanic fuzzy
neural networks. In: Proceedings of the
1999 America Control Conference (Cat.
No 99CH36251. Vol. 4. IEEE; 1999.
pp. 2310-2314. DOI: 10.1109/
ACC.1999.786445

[33] Schäfer P, Caspari A, Kleinhans K,
Mhamdi A, Mitsos A. Reduced dynamic
modeling approach for rectification
columns based on compartmentalization
and artificial neural networks. AICHE
Journal. 2019;**65**(5):e16568.
DOI: 10.1002/aic.16568

[34] Linhart O, Gela D, Rodina M,
Kocour M. Optimization of artificial
propagation in European catfish, Silurus
glanis L. Aquaculture. 2004;**235**(1–4):
619-632. DOI: 10.1016/j.
aquaculture.2003.11.031

Chapter 9

Heat Integration of Reactive Divided Wall Distillation Column

Rajeev Kumar Dohare and Parvez Ansari

Abstract

In this chapter, the esterification reaction of methyl acetate with methanol and acetic acid is proposed by the new column configuration that is a reactive divided wall distillation column (RDWC). The heat integration of the proposed column configuration is studied and found that heat integration techniques are efficient to save energy up to 14.23% in comparison to the conventional reactive divided wall distillation column.

Keywords: methyl acetate, reactive dividing wall column, simulation, heat integration

1. Introduction

Distillation Technique is a chief process division extensively used in chemical manufacturing, and the growth of distillation column strategy has involved more and more consideration in the latest ages [1, 2]. Reactive distillation (RD) is a mixture of synthetic reaction and partition simultaneously. Catalytic distillation (CD) is a Reactive distillation process in which chemical reactions happen in a solid catalyst. The blend of the split process with catalytic reactions in a catalytic distillation column has many benefits, for example, improved conversion for stability and expected improvement of item selectivity because of the removal of the yields over in situ partings. In accumulation, the produced heat in the reaction might practice for distillation. Thus, the investment and operative expenses can be reduced [3, 4]. The reactive distillation process has been a part of importance most recent 20 years. The mixture of reaction and separation in a single unit is a substitute to traditional distillation, which incorporates reaction, and division in the number of units consequently expands the investment cost of the plant. The three significant useful areas of the reactive distillation column are:

1. Distillation column comprises a reactive unit, which prompts the transformation of reactants into items

2. It works on the partition in the section by changing the part volatilities

3. Intensifications the selectivity of the item [5].

To feat the likelihood of reactive distillation, procedures have been established for initial process design. Two major methods happen for the group of substitutes for a given reaction-parting problem: mathematical optimization and graphically based conceptual design methods. Mathematical optimization methods are generally very powerful for generating and evaluating design alternatives [6]. Reactive distillation technology is an encouraging process that can rise reaction change, overcome energy feeding, and recover selectivity and investment yield [7]. In previous years, Reactive distillation was widely inspected due to the projecting benefits, and excessive successes have been gained in positions of dynamic control and development strategy [8–12]. The results of the simulation proved that this novel technology is economical on the process charges and easily controls the purity of the product with the only use of a temperature control loop, which shows the potentials of the heat-integrated reactive distillation methodology. To complete additional energy savings, scientists have focused on the project of DWC to superior distillation structures, for example, reactive divided wall distillation, extractive distillation, or azeotropic distillation [10, 13, 14]. Reactive dividing wall column (RDWC) has newly involved countless importance since it has the rewards of together RD and DWC. A technique for the theoretical strategy of the RDWC created on the smallest vapor stream process planned the rate-based modeling method for RDWC [15]. Therefore, it is necessary to study the Heat integrated reactive distillation performances of reaction systems with different characteristics.

ASPEN PLUS simulation provides the benefits in covering steady-state to dynamic simulation for safety analysis and control process. The physical properties methods are required during the Aspen model, to calculate enthalpy, density viscosity, heat capacity, etc. ASPEN PLUS simulator has been used for physical chemistry, chemical thermodynamics, mass and energy balances, chemical reaction engineering, unit operations, and process design and control. It uses an inbuilt numerical model equation to stabilized the process performance. The perfect showing of thermodynamic properties is mainly significant in the parting of non-ideal mixtures and ASPEN PLUS has big information of retreated factors. Methanol (MeOH) and acetic acid (HAc) are essential raw materials in polyvinyl alcohol plants, and they could produce from Methyl Acetate (MeAc) hydrolysis process. Therefore, this process is considering for heat integration purposes.

1.1 Dividing wall column (DWC)

The petrochemical and chemical divisions are the major manufacturing power clients, representing generally 10% of overall global energy interest and 7% of worldwide greenhouse gases (GHG) outflows. In the chemical process industry, roughly distillation processes utilize 40% of absolute energy [16]. In the distillation procedure, high temperature is utilized, for example, an isolating means. Heat is provided to the lower section of the reboiler to vanish a fluid saturation at a more temperature and it decreases at less temperature while melting in the condenser at the upper section of the distillation column. Hence, the situation is extremely unproductive in the utilization of power. In the 1970s and 1980s, the start of oil emergencies, the power costs turned into the central point in column rate and made a resolve to discover to decrease the energy requirements of distillation. Subsequently, in the new distillation process, the essential objective is process strategies in distillation systems is that how to cut the power utilization. Different strategies have been used to utilized to make the process of distillation more energy effective and extra economical like divided wall columns (DWC), heat integrated distillation columns (HIDiC), and thermally coupled distillation columns (Petlyuk column).

In **Figure 1**, Wright's patent the divided wall column (DWC) in 1949. DWC can save both energy requirements and economic expenses related to conventional distillations. The energy utilization decreases about 20–30% associate with another conventional distillation column [17, 18].

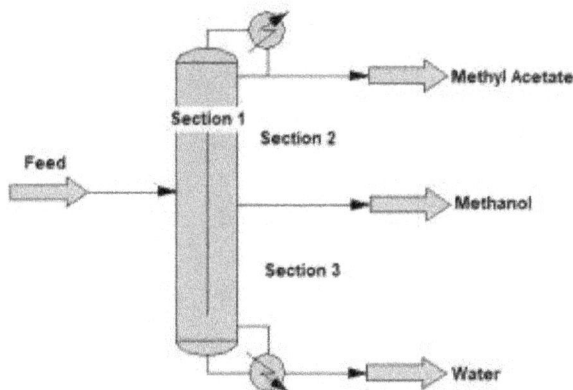

Figure 1.
Divided wall column.

The Divided wall column contains more than two distinctive split process units into single and more than one vertical section in the middle area. Dividing walls also differentiate a single column into two sections: a pre-fractionator area and the main column. It also used the condenser and reboiler at the top and bottom respectively [19].

Advantages and disadvantages of divided wall columns

1. Lower capital investment

2. Reduced energy requirements

3. High purity for all products

4. Less construction volume

A divided wall column might be offered the potential for decreasing both economic and energy prices, the dividing wall columns have main disadvantages. They are:

1. Higher columns dues to the increased number of theoretical stages.

2. Due to the higher number of theoretical stages, the increase in pressure drops.

3. Operating pressure is available only once.

1.2 Reactive distillation column (RDC)

Many numerical problems arise in the modeling, design, and optimization of the RDC, which results in simpler and intensified processes with fewer recycle streams, and decreasing waste handling reflects lower investments and operating costs. RDC offers an advanced reaction rate and selectivity; stops the performance of azeotrope, less energy intake, and solvent treatment. Despite all these benefits, the RDC has partial commercial applications; it is because of the control performance and the complexity in the operation of the RDC. For modeling, we have supposed that it operates in adiabatic conditions with the liquid phase. There is no vapor hold-up

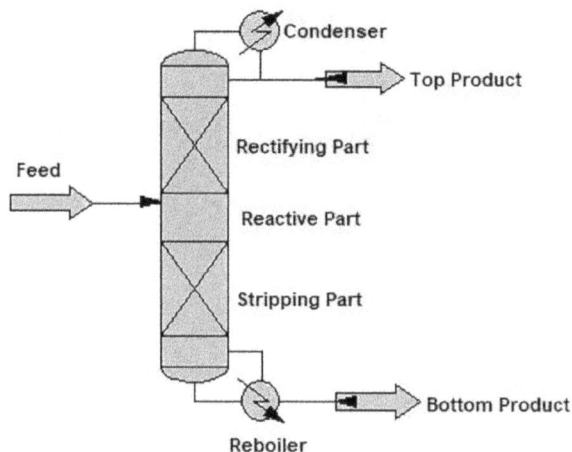

Figure 2.
Reactive distillation column.

in any stage of the distillation column (DC). No hydrodynamic effects have been considered escaping the modeling difficulties [20]. **Figure 2** is an actual sketch of the reactive distillation column.

1.3 Reactive dividing wall distillation column (RDWC)

A reactive dividing-wall column (RDWC) incorporates a reactor and a separator in a single distillation column. The multiple products, non-reacting components, or excessive reagents can be isolated in such a column, that reactive systems have. Because of the strong corporation among control loops, control engineers have a provoking position to control RDWC. Up until now, the investigation of reactive distillation in one divided wall column is scant, particularly for the control. The reason for this work is to consolidate the advantages of reactive distillation with DWC to deliver MeAc and afterward examine the design and control of an RDWC with an exceptional focus on the foundation of control structures. Initially, a subjective connection between the process flow sheet and phase equilibria is set up, and then an RDWC flow sheet is set up.

The reactive distillation column (RDC) and dividing wall column (DWC) both are genuine instances of process heat intensification. Uncertainty reactive distillation and DWC have combined, a reactive divided wall column (RDWC) has been produced. RDWC has an extremely integrated arrangement that contains one condenser, one reboiler, reactive zones, a pre-fractionator, and the main column together in a single distillation setup. The synthesis of Methyl Acetate has been chosen as a test reaction for heat integration purposes.

The synthesis of Methyl Acetate and its reverse reaction are given below.

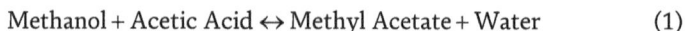

$$\text{Methanol} + \text{Acetic Acid} \leftrightarrow \text{Methyl Acetate} + \text{Water} \qquad (1)$$

In reactive distillation, it is likely to get more conversion by continuously removing the products from the reaction section. Products have been removed from the lower part of a reactive distillation column and isolated into the distillation column. By joining the reactive distillation column and separation column into the single column, which turns into the reactive divided wall column with side product methanol stream, also the residence period of methanol with acetic acid and water in the sump has come to a minimum level.

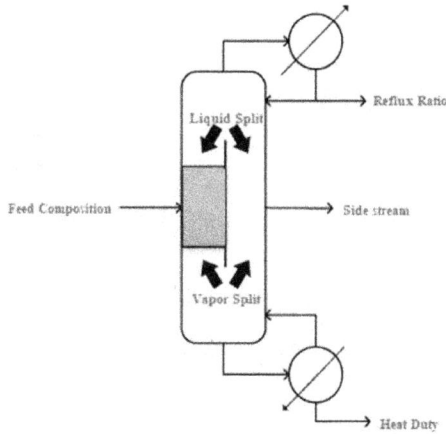

Figure 3.
Reactive dividing wall column.

It has achieved greater attention in the chemical industry for the separation process and saves both energy and capital cost. The RDWC technology has not been confined up to ternary separation only but it can also carry out azeotropic separations. The feasibility of the RDWC in the industry depends upon the thermodynamic properties, the composition of the stream that has separated, and the product requirements. **Figure 3** shows the actual pictorial diagram of the RDWC. Although, **Figure 4** shows the simulation sheet of the two-column reactive divided wall distillation column.

1.4 Heat integration of reactive dividing wall column

The main approach to improve the energy efficiency of the distillation system is by providing heat integration technology [21]; vapor recompression (VRC) and internally heat-integrated distillation column (HIDiC) [22] are two popular techniques for the same. The energy demands and expenses are expanding due to joined hazards, global warming, and the improved requirement upon lubricant introduced from electorally insecure quantities of the sphere have caused in the importance of the thermodynamic efficiency of recent engineering progress for

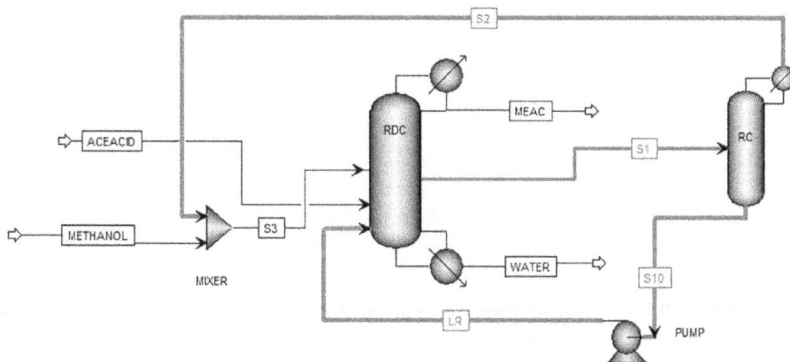

Figure 4.
Flowsheet of RDWC.

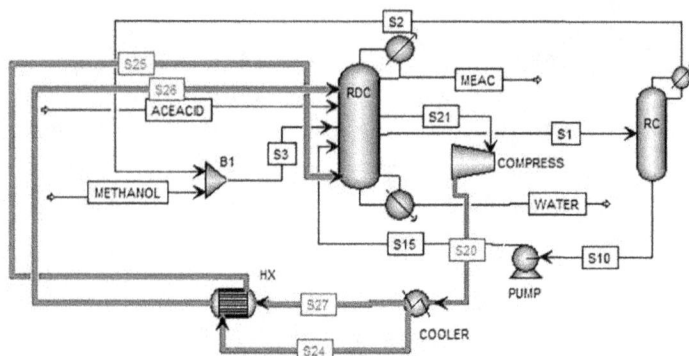

Figure 5.
Flowsheet of heat integrated reactive dividing wall column.

improving. Expanding power efficiency in compound routes not individual offers cost-effective profits and turns into the decreasing radiations resultant from the development activity. The distillation technology is maybe the important significant and extensively used removable technique in the world today about 95% of all liquid split in the synthetic production industry used. Regardless of its evident significance, in general, most of the thermodynamic efficiency of a conventional distillation is about 5–20% [23]. The concept of process intensification has presented to the distillation development in 1970 for further development of energy efficiency [24]. Nevertheless, the energy efficiency of RDWC is not constantly high since the entire heat can only be added to the bottom reboiler at the maximum temperature and impassive from the top condenser at the final temperature. The heat integration technology such as VRC the exciting energy intake has been reduced [25, 26]. The conventional VRC and side VRC techniques are combined into one RDWC; to deliver the intensified heat integrated technique of VRC that are associated with the RWDC structure. Into intensified heat integrated structures, the unoriginal VRC has further divided between the top and bottom of the RDWC to improve overheat vapor through the heat flow as greatly as possible, into more temperature, compressed fluid from the downward reboiler has recycled into the adjacent fluid to vaporize in a transitional reboiler (IR) [27]. The fundamental thought of heat integration technology is that the high-temperature route streams that exchanged heat with cold route streams, which affect in financial use of assets. Subsequently, numerous sorts of exploration on heat-integrated distillation column (HIDiC) has been done in the past few years to research its achievability and reasonableness in certifiable applications as shown in **Figure 5**. Distillation columns such as the Petlyuk column, divided-wall column, heat pump assisted column, adiabatic distillation column, ideal HIDiC (i- HIDiC), and many more this investigation has led to the formation of different technology. The HIDiC structure has until not accepted by many industries then later small scale trials are accepted by the New Energy and Industrial Technology Development Organization (NEDO), Japan; a combined organization between NEDO and TERI (The Energy & Resources Institute), India undertakings more projects in the field of process intensification technology at a conference at New Delhi in January 2017 [28]. This tends to an original thought of process integration by joining reactive dividing wall column and ethics of heat integrated distillation column i.e. R-HIDiC that accepted the synthesis of methyl acetate with significantly lesser energy consumption. The mixture of methyl acetate employing a reactive dividing wall column system is the most effective and main application of process intensification [29]. Nevertheless, the greatly needed attention to decrement the energy feeding and intensification the efficiency of the current reactive dividing wall column has considered in this research.

2. Methodology

2.1 Column configuration

The method used for the simulation is the UNIQAC method, RadFrac model is used for the design of RDWC and Hi-RDWC in ASPEN simulation. The UNIFAC method is used for the estimate of activity coefficients calculated on the idea that a fluid mixture might be measured as a result of the structural elements after which the particles are produced slightly than a result of the particles themselves. The RadFrac model is the chief partition block option in Aspen Plus. The block option can execute sizing, simulation, rating of the tray, and packed columns. The ideal requirement, completed in the Format form, needs full conditions of column structure, feed, product, and any side streams. The feed material was methanol and acetic acid. The detailed operating parameters are shown in **Table 1**.

2.2 Intensified heat integration configurations

The standard VRC assisted RDWC configuration (VRC-RDWC) presented in **Figure 6**, the overhead vapor has compacted to a surpassing temperature to turns into vapor the lowermost fluid, and formerly converts the saturated condensed fluid into the reboiler, and the condensed fluid must be low to the topmost pressure through the throttle valve (TV) that returning at the RDWC at the upper section. Now this method, in the isentropic compressor the less amount of saturated vapor was condensed and less amount of saturated fluid will flash into the throttle valve, so a heater was essential to preheat the overhead vapor and the flew vapor condensed totally by the use of cooler. While the compression ratio (CR) is

Sr. no	Parameters	Value (unit)
1	Feed-Methanol	100 mol/hr
2	Feed–Acetic Acid	80 mol/hr
3	Pressure	1.013 bar
4	Reflux Ratio (RDC)	6.7
5	Distillate Rate (RDC)	80 mol/hr
6	Side Rate (RC)	20 mol/hr
7	Bottom Rate (RC)	80 mol/hr
8	Feed Tray location (a) Methanol	10
9	Feed Tray location (b) Acetic Acid	2
10	Number of Stages	36
11	Number of Stages RDC	25
12	Number of Stages RC	11

Table 1.
Operating parameters.

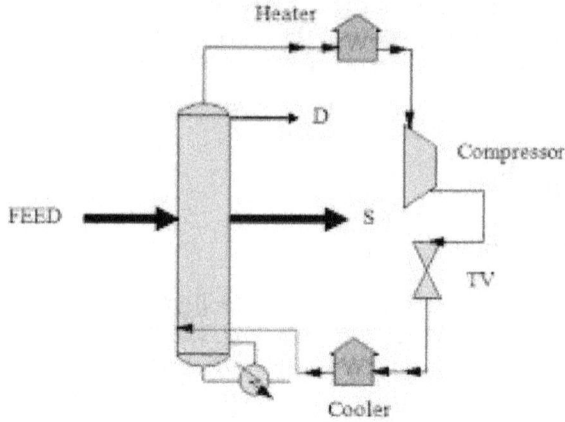

Figure 6.
Schematic diagram of vapor recompression RDWC (VRC-RDWC).

insignificant, the heater and cooler have been discounted due to the heat duties, then below large CR, these heat duties will convert huge and straight shrink the power effectiveness of VRC knowingly. The unused heat created by the VRC has recovered by intensifying the process of the heat integration technique between the VRC and RDWC (**Figure 5**).

3. Results and discussion

3.1 Temperature and composition profile

The temperature difference between the overhead and the bottom product is high i.e. 98°C. Therefore, the compression ratio (CR) has been regulated to meet the heat transfer needed in VRC-RDWC structures as shown in **Figure 5**. The total number of stages used in both columns including i.e. RDC and Rectifying column (RC)) are 36. The composition profile of the RDC column is shown in **Figure 7**. As per the physical phenomena that the excess amount of reactant and products should be at the output side. Although, the full consumption of the limiting reactant. The graph shows a similar response as per the process phenomena. The temperature of the column continuously increases with the increase in the number of the stages as

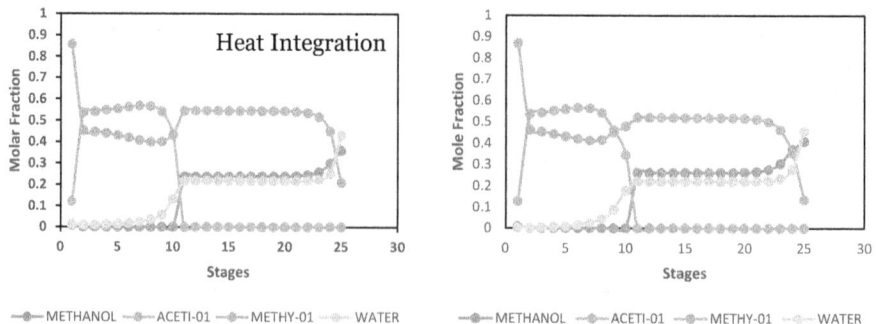

Figure 7.
Composition profile of RDC column (heat integration and without heat integration).

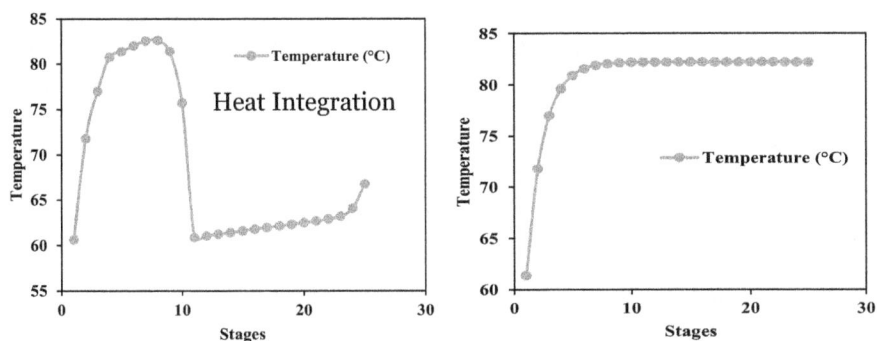

Figure 8.
Temperature profile of RDC column (heat integration and without heat integration).

Results	RDWC (2 column)	Hi-RDWC (2 column)	Energy saving (%) (2 column)
Condenser duty (kW)	−1.386	−1.358	0.020
Cooler duty (kW)	0	0.0936	0
Reboiler duty (kW)	0.6667	0.5718	14.23
Heater duty (kW)	0	0	0
Compressor duty (kW)	0	2.302e-05	0

Table 2.
Results of heat integrated RDWC (2 column).

shown in **Figure 8.** The maximum has risen the temperature is 85°C but with the implementation of the heat integration technique, the temperature is decreasing at about 65°C. There are no changes in composition profile by the use of heat integration technique as shown in **Figure 7.**

3.2 Heat integration of two column RDWC

To analyze the heat recovery in two columns design, a rigorous simulation has performed for the heat integrated simulation flowsheet. The Aspen simulation flowsheet of the RDWC is shown in **Figure 4.** Therefore, an enormous amount of condensed fluid flashed in the throttle valve (TV) from the compression ratio. The cooler with a heat duty of 0.020 kW condenses the flashed liquid. The reboiler duty of RDWC is 0.6667 kW after the Heat Integration of RDWC the reboiler duty was decreased by 0.5718 kW. However, the energy-saving of the intensified configuration technology is significant in comparison to the conventional column. The heat-integrated data of the divided wall distillation column is given in **Table 2.**

4. Conclusion

In this book chapter, Aspen Plus software is used to simulate the process of producing methyl acetate. The new technology used in this research was reactive dividing wall distillation technology. The position of methanol and acetic acid feed stream is set to be on the Reactive Distillation column (RDC) at 10 and 2 respectively. It is concluded that after the VRC-RDWC heat integration technique the reboiler duty is reduced from 0.6667 to 0.5718 kW and the condenser duty is reduced from −1.386 to −1.358 kW in the case of two-column configuration Therefore, it is observed after the integration the heat load on the reboiler is reduced up to 14.23% and condenser duty is to reduce up to 0.020% in case of two-column configuration. Results also showed that all products compositions could be kept at desired purity against feed disorder.

Acknowledgements

I am thankful to the department of chemical engineering to provide the research facilities and infrastructure to conduct this work. I want to extend my thanks to DST, SERB Delhi to provide the financial support for this research work.

Conflict of interest

There is no interest of conflict. The used data is properly cited and acknowledged here.

Author details

Rajeev Kumar Dohare* and Parvez Ansari
Department of Chemical Engineering, Malaviya National Institute of Technology, Jaipur, Rajasthan, India

*Address all correspondence to: rajeevdohare@gmail.com

IntechOpen

References

[1] Grossmann IE, Martín M. Energy and water optimization in biofuel plants. Chinese Journal of Chemical Engineering. 2010;**18**(6):914-922. DOI: 10.1016/S1004-9541(09)60148-8

[2] Fang J, Zhao H, Qi J, Li C, Qi J, Guo J. Energy conserving effects of dividing wall column. Chinese Journal of Chemical Engineering. 2015;**23**(6):934-940. DOI: 10.1016/j.cjche.2014.08.009

[3] Doherty G, Buzad MF. "Reactive distillation by design," Transactions. Institute of Chemical Engineers. 1992;**70**(A):448-458.

[4] Sundmacher K, Hoffmann U. "Multicomponent mass and energy transport on different length scales in a packed reactive distillation column for heterogeneously catalysed fuel ether production." Chemical Engineering Science. 1994;**49**(24 PART A):4443-4464. DOI: 10.1016/ S0009-2509(05) 80032-6.

[5] Giwa KSA. "Development of Dynamic Models for a Reactive Packed Distillation Column." International Journal of Engineering. 2012;**6**(3):118-128.

[6] Ciric AR, Gu D. Synthesis of nonequilibrium reactive distillation processes by MINLP optimization. AICHE Journal. 1994;**40**(9):1479-1487. DOI: 10.1002/aic.690400907

[7] Yang B, Wu J, Zhao G, Wang H, Lu S. Multiplicity analysis in reactive distillation column using ASPEN PLUS. Chinese Journal of Chemical Engineering. 2006;**14**(3):301-308. DOI: 10.1016/S1004-9541(06)60075-X

[8] Dimian AC, Bildea CS, Omota F, Kiss AA. Innovative process for fatty acid esters by dual reactive distillation. Computers and Chemical Engineering. 2009;**33**(3):743-750. DOI: 10.1016/j.compchemeng.2008.09.020

[9] Li S, Huang D. Simulation and analysis on multiple steady states of an industrial acetic acid dehydration system. Chinese Journal of Chemical Engineering. 2011;**19**(6):983-989. DOI: 10.1016/S1004-9541(11)60081-5

[10] Bildea CS, Gyorgy R, Sánchez-Ramírez E, Quiroz-Ramírez JJ, Segovia-Hernandez JG, Kiss AA. Optimal design and plantwide control of novel processes for di-n-pentyl ether production. Journal of Chemical Technology and Biotechnology. 2015;**90**(6):992-1001. DOI: 10.1002/jctb.4683

[11] Luyben WL. Distillation Design and Control Using Aspen Simulation. John Wiley: Wiley-Interscience A John Wiley & Sons, Inc.; 2013

[12] Gao X, Li X, Li H. Hydrolysis of methyl acetate via catalytic distillation: Simulation and design of new technological process. Chemical Engineering and Processing Process Intensification. 2010;**49**(12):1267-1276. DOI: 10.1016/j.cep.2010.09.015

[13] Kiss AA, Pragt JJ, van Strien CJG. Reactive dividing-wall columns-how to get more with less resources? Chemical Engineering Communications. 2009; **196**(11):1366-1374. DOI: 10.1080/00986440902935507

[14] Sun L, Wang Q, Li L, Zhai J. Design and control of extractive dividing wall column for separating benzene/cyclohexane mixtures. Industrial and Engineering Chemistry Research. 2014;**53**(19):8120-8131

[15] Wang SJ, Wong DSH, Yu SW. Design and control of transesterification reactive distillation with thermal coupling. Computers and Chemical Engineering. 2008;**32**(12):3030-3037. DOI: 10.1016/j.compchemeng. 2008.04.001

[16] Jana AK. Heat integrated distillation operation. Applied Energy. 2010;**87**(5): 1477-1494. DOI: 10.1016/j.apenergy. 2009.10.014

[17] Bandaru, Kiran AKJ. Introducing vapor recompression mechanism in heat-integrated distillation column: Impact of internal energy driven intermediate and bottom reboiler. (Wiley Online Library) AIChE Journal. 2014;**61**:118-131. DOI: 10.1002/aic.14620

[18] Jana AK. Advances in heat pump assisted distillation column: A review. Energy Conversion and Management. 2014;**77**:287-297. DOI: 10.1016/J. ENCONMAN.2013.09.055

[19] Dejanović I, Matijašević L, Olujić Ž. Dividing wall column-A breakthrough towards sustainable distilling. Chemical Engineering and Processing Process Intensification. 2010;**49**(6):559-580. DOI: 10.1016/j.cep.2010.04.001

[20] Calzon-McConville CJ, Rosales-Zamora MB, Hernández S, Segovia-Hernández JG, Rico-Ramírez V. Design and optimization of thermally coupled distillation schemes for the separation of multicomponent mixtures. Industrial and Engineering Chemistry Research. 2006;**45**(2): 724-732. DOI: 10.1021/ie050961s

[21] Bodnarchuk MS, Heyes DM, Breakspear A, Chahine S, Edwards S, Dini D. Response of calcium carbonate nanoparticles in hydrophobic solvent to pressure, temperature, and water. Journal of Physical Chemistry C. 2015;**119**(29):16879-16888. DOI: 10.1021/acs.jpcc.5b00364

[22] Keshwani DR, Cheng JJ. Switchgrass for bioethanol and other value-added applications: A review. Bioresource Technology. 2009;**100**(4):1515-1523. DOI: 10.1016/j.biortech.2008.09.035

[23] Nguyen TD. Conceptual design, simulation and experimental validation of divided wall column: application for nonreactive and reactive mixture [thesis]. Toulouse, France: Université de Toulouse; 2015. p. 179

[24] An D, Cai W, Xia M, Zhang X, Wang F. Design and control of reactive dividing-wall column for the production of methyl acetate. Chemical Engineering and Processing Process Intensification. 2015;**92**:45-60. DOI: 10.1016/j.cep.2015.03.026

[25] Mascia M, Ferrara F, Vacca A, Tola G, Errico M. Design of heat integrated distillation systems for a light ends separation plant. Applied Thermal Engineering. 2007;**27**(7):1205-1211. DOI: 10.1016/j.applthermaleng. 2006.02.045

[26] Dohare K. Simulated heat integration study of reactive distillation column for ethanol synthesis. Iranian journal of chemistry and chemical engineering. 2019;**38**(4):183-191

[27] Annakou O, Mizsey P. Rigorous investigation of heat pump assisted distillation. Heat Recovery Systems and CHP. Apr. 1995;**15**(3):241-247. DOI: 10.1016/0890-4332(95)90008-X

[28] Ahmed SA, Ahmad SA. Modelling of heat integrated reactive distillation column (r-HIDiC): Simulation studies of MTBE synthesis. Indian Journal of Chemical Technology. 2020;**27**:210-218

[29] Luyben WL, Yu CC. Design of MTBE and ETBE reactive distillation columns. Reactive Distillation Design and Control. 2009;**29**:213-237. DOI: 10.1002/9780470377741.ch9

Centralized and Decentralized Control System for Reactive Distillation Diphenyl Carbonate Process

Shirish Prakash Bandsode and Chandra Shekar Besta

Abstract

Reactive distillation (RD), a process-intensified technique, involves the integration of reaction and separation in a single unit. High non-linearities associated with the reactive distillation process constrict the control degrees of freedom and set the key challenge in the design of a robust control system. In this chapter, reactive distillation diphenyl carbonate (RD-DPC) design is optimized, and a decentralized as well as centralized feedback control configuration is applied to carry out the control studies. To execute the control scheme, a dynamic model of RD-DPC process is developed using Aspen Dynamic and interfaced with MATLAB Simulink for online control implementation. A comparative multi-loop feedback controller control performance study is done for different transfer function models obtained by using analytical- and optimization-based process identification techniques. The controller parameters obtained from the simple internal model control (SIMC) tuning relations for decentralized controller and Tanttu & Lieslehto (TL) tuning relations for centralized controller are applied to (i) the linear transfer function model and (ii) non-linear plant model. Set-point tracking, load rejection studies and robust stability analysis are carried out to compare the performance of different models and to investigate the controller performance of the non-linear model.

Keywords: reactive distillation, diphenyl carbonate (DPC), decentralized controller, centralized controller, robustness, non-linear model

1. Introduction

Polycarbonates, containing carbonate groups in their chemical structures, are an important group of thermoplastic polymers. Diphenyl carbonate (DPC), an acyclic carbonate ester, is a monomer in the production of polycarbonate polymers. The production of DPC is carried out by the transesterification reaction between dimethyl carbonate (DMC) and phenyl acetate (PA). The reactive distillation process, involving the integration of reaction and separation in one place, is usually associated with high non-linearities. The interaction of reaction and separation, responsible for the occurrence of multiple steady states, sets a challenge in designing a robust controller. Furthermore, the high non-linearity and dynamic interactions cannot be effectively controlled by single-input single-output (SISO) controller and hence urges for multi-input multi-output (MIMO) controller.

In this work, RD–DPC process model is simulated using Aspen Dynamic V11. The transfer function model and controller development are performed using MATLAB 2019b Simulink Control system and custom proportional-integral-derivative (PID) coding. An online control environment is created by interfacing Aspen Dynamic with MATLAB Simulink via AM System block and similarly linking the centralized controller to MATLAB Simulink via S-function block.

This chapter reflects the designing of RD-DPC two-column indirect sequence and a control system for maintaining the molar purity of DPC and methyl acetate (MA) greater than 99%. The chapter also shows a comparative study between control performance of decentralized and centralized feedback controllers.

2. RD-DPC multivariable process

DPC is produced by reacting phenyl acetate (PA) and dimethyl carbonate (DMC) in a reactive distillation column. The involved reactions and the corresponding reaction rates are mentioned subsequently (Eqs. (1)–(3)). The reaction kinetic constants for the forward and backward reactions are taken from the work done by Cheng et al. [1]. There is a rectification and a reaction zone in the RD column, as shown in **Figure 1**. Column design specifications and additional parameters are reported in **Table 1**. Although high-purity DPC is obtained at the bottoms of the RD column, the purity of methyl acetate (MA) obtained at the distillate of the RD column is low. To obtain MA at the desired purity, we have to use another separation column, thus reactive distillation plus non-reactive distillation.

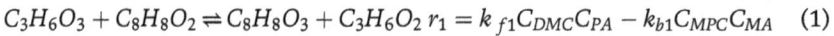

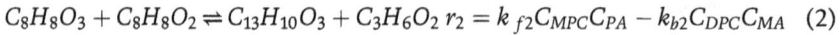

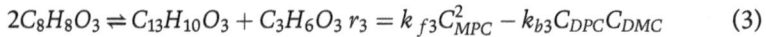

$$C_3H_6O_3 + C_8H_8O_2 \rightleftharpoons C_8H_8O_3 + C_3H_6O_2 \ r_1 = k_{f1}C_{DMC}C_{PA} - k_{b1}C_{MPC}C_{MA} \quad (1)$$

$$C_8H_8O_3 + C_8H_8O_2 \rightleftharpoons C_{13}H_{10}O_3 + C_3H_6O_2 \ r_2 = k_{f2}C_{MPC}C_{PA} - k_{b2}C_{DPC}C_{MA} \quad (2)$$

$$2C_8H_8O_3 \rightleftharpoons C_{13}H_{10}O_3 + C_3H_6O_3 \ r_3 = k_{f3}C_{MPC}^2 - k_{b3}C_{DPC}C_{DMC} \quad (3)$$

Aspen Plus/Dynamics is used to design and simulate the RD-DPC indirect sequence. The steady-state simulation results are shown in **Table 2**. In terms of the model validation, the required data are taken from the original case study [1, 2].

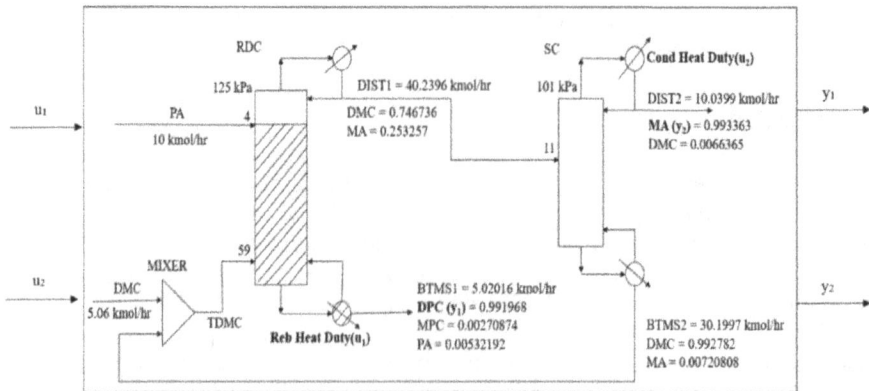

Figure 1.
The conventional RD-DPC process. [y(s) = Gp(s) u(s)].

Sr. No	Parameters	RD	SC
1	Total number of stages	65	25
2	Number of reactive stages	4–65	—
3	Feed stage	4, 59	11
4	Reflux ratio	1.25	2.48
5	Operating pressure (kPa)	125	101
6	Stage pressure drop (kPa)	0.625	0.63
7	Tray holdup	0.097	—
8	Column diameter (m)	1.022	0.642
9	Height of column (m)	32.5	12.5
10	Condenser duty (kW)	−839.704	−296.229
11	Reboiler duty (kW)	892.996	297.697

Table 1.
Design specifications and parameters.

Description	PA	DMC	TDMC	BTMS1	DIST1	BTMS2	DIST2
T (C)	204.54	103.5	95.5	330.1	84.7	94.2	57.1
P (kPa)	127	152	163	165.32	125	116.12	101
Mole Flow (kmol/hr)	10	5.06	35.26	5.02	40.24	30.2	10.04
Mole fraction							
MA	0	0	0.00618	0	0.26	0.0072	0.994
PA	1	0	0	0.0054	0	0	0
DMC	0	1	0.994	0	0.74	0.993	0.0067
DPC	0	0	0	0.992	0	0	0
MPC	0	0	0	0.0028	0	0	0

Table 2.
Steady-state simulation results.

This case study is close to a real RD-DPC process in terms of sizing, as reported by the researchers; thus, the validation data can be regarded as industrial data as well. Here, the liquid mole fraction profile is chosen for validation. **Figure 2** depicts the performance of simulating the RD-DPC liquid mole fraction profile and validation results. Aspen model managed to present accurate results by validating the data having an R^2 value of ≈ 0.9.

3. Open-loop dynamic analysis

In general, reactive distillation is usually associated with the occurrence of multiple steady states [3]. Occurrence of multiplicity is a consequence of the high non-linearities associated with the RD process. The cause of multiplicity is connected with the presence of multiple reactions, heat of reaction, and the crossing of non-reactive distillation boundary via reaction. Multiplicity in the form of input multiplicity or output multiplicity exists in the RD process. Input multiplicity alters

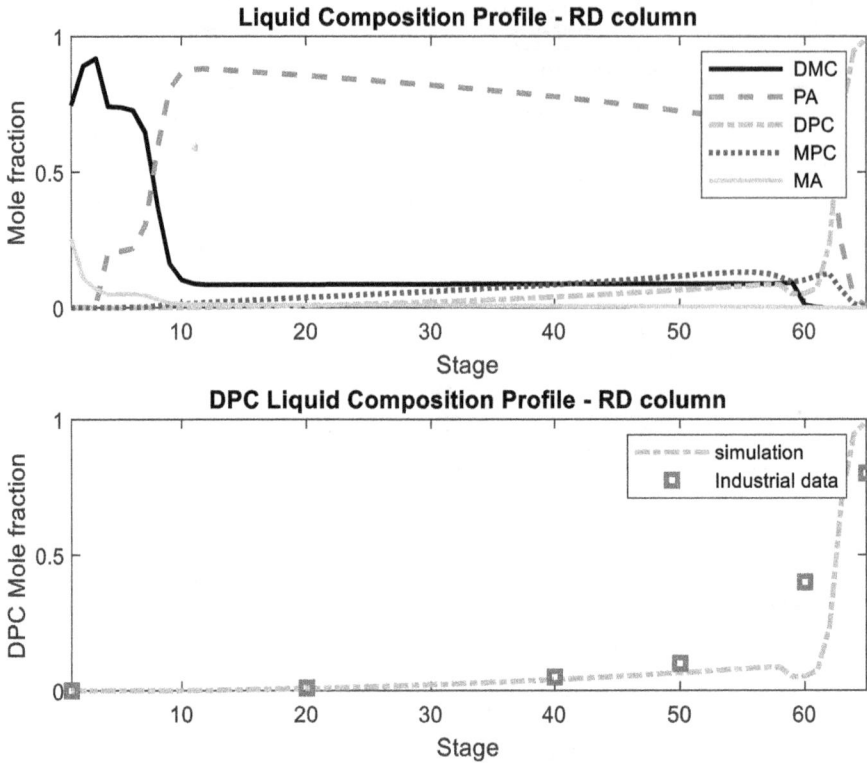

Figure 2.
Aspen model liquid mole fraction profile compared with industry data.

the selection of controlled variables, whereas output multiplicity affects the choices of control structure and the operating range [4]. In open-loop analysis, a series of step changes were applied to the manipulated variable (u_1 and u_2) in order to check for the presence of multiplicity and to set up the operating range.

4. RD-DPC control system design

To analyze the control performance of the RD-DPC process, a two-input two-output (TITO) multivariable system with time delay is considered [5]. $G_{P(s)}$ represent the process transfer function. Similarly, $G_{c-D(s)}$ and $G_{c-C(s)}$ represent the decentralized controller [6–10] and centralized controller [11–15], respectively. Controller output and process output are represented by u_i and y_i, respectively.

$$G_P(s) = \begin{bmatrix} g_{P,11}(s) & g_{P,12}(s) \\ g_{P,21}(s) & g_{P,22}(s) \end{bmatrix} \tag{4}$$

$$G_{c-D}(s) = \begin{bmatrix} g_{c,11}(s) & - \\ - & g_{c,22}(s) \end{bmatrix} \tag{5}$$

$$G_{c-C}(s) = \begin{bmatrix} g_{c,11}(s) & g_{c,12}(s) \\ g_{c,21}(s) & g_{c,22}(s) \end{bmatrix} \tag{6}$$

For controller settings, the SIMC tuning relations [16] are used to design the decentralized controller. Similarly, Tanttu & Lieslehto (TL) [17] tuning relations are used for calculating centralized controller settings. The controller performance is assessed by considering the setpoint tracking, settling time, and disturbance rejection tests. The controller's ability to properly move to another purity level is assessed in the grade transition test. The disturbances variables are the feed flow rate of PA to the RD column (d_1) and the reboiler heat duty of the separation column (d_2). These variables are typically more exposed to disturbances since they are originated from outside of the system. The controller disturbance rejection potential is evaluated by doubling the amount of disturbance to the standard reported in the industry. The controller's performance is evaluated by employing integral square error (ISE).

5. Results and discussion

5.1 Open-loop analysis

In open-loop analysis, the existence of multiple steady states is observed, and the operating window for the variables is set. This section is divided into two parts. **(i) Step changes in RD reboiler heat duty:** A series of step changes were applied to the reboiler heat duty of the RD column (the manipulated variable for controlling the molar purity of DPC), to fix the operating range. A step change of $\pm1.5\%$ was applied to the reboiler heat duty, and the corresponding dynamic response was observed for the controlled variables y_1 and y_2. It was found that the column sets at other steady state and the desired molar purities are not achieved. Thus, reduced step changes were applied to u_1. For a series of step changes of $\pm1\%$ applied to u_1, the desired molar purities were not achieved. Similarly, for step changes of $\pm0.75\%$ to u_1, the desired molar purities of DPC and MA are obtained. Thus, the manipulated variable for the RD column is operated between 886.237 kW and 899.631 kW. **(ii) Step changes in separation column condenser heat duty:** Here, step changes were applied to the condenser heat duty of the separation column (the manipulated variable for controlling the molar purity of methyl acetate (MA)), in order to fix the operating region. For step changes of $\pm3\%$ and $\pm 2\%$ to the condenser heat duty of separation column, the desired molar purities are not achieved and the process sets at another steady state, indicating the presence of multiplicity. Similarly, a step change of $\pm1\%$ is applied to u_2, **Figure 3**, and the responses in y_1 and y_2 are observed. It was observed that the desired molar purities are obtained, and thus the manipulated variable for separation column is operated between -293.240 kW and $- 299.164$ kW.

5.2 RD-DPC model identification

This section describes how the model Identification for RD–DPC process is carried out [18]. When the matching process employs optimization, a model prediction is aligned with the measured values with the use of a solver. Eq. (A1) has variables $y(t)$ and $u(t)$ and two unknown parameters K_p and τ_p. These variables may be adjusted to match the data. The solver often minimizes a measure of the alignment, such as a sum of the squared errors or sum of absolute errors. The optimization solver used in excel is "generalized reduced gradient (GRG) non-linear." Here, we have two manipulated variables u_1 and u_2. When we give a step change in u_1, we observe the response in y_1 and y_2, respectively, and similarly a step input to u_2 gives a response in y_1 and y_2. So, in total we have four data sets, u_1-y_1, u_1-y_2, u_2-y_1, and u_2-y_2.

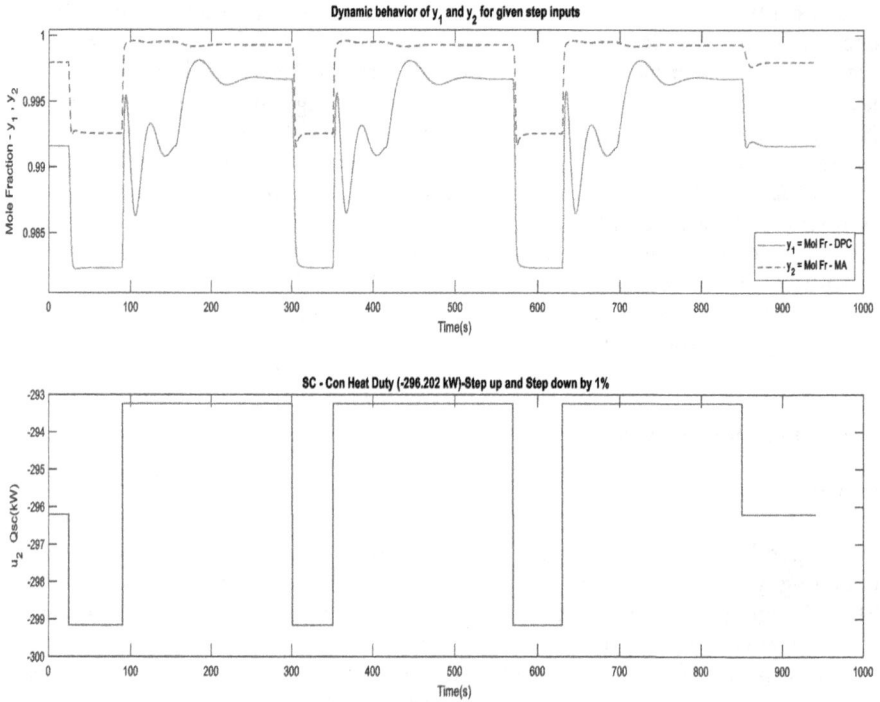

Figure 3.
Open-loop dynamic behavior for NL-RD-DPC process of interaction (y_1) and response (y_2) for a given step change (1%) in condenser heat duty (kW) of SC column (u_2).

For the obtained datasets, the variables when adjusted give us four models – g_{11}, g_{21}, g_{12}, and g_{22}, respectively. Similarly, the optimization solver "SciPy.Optimize.Mini-mize" function in Python, changes the unknown parameters of Eqs. (A2) and (A3) to best match the data at specified time points. The sum-of-squared errors and the obtained values of the unknown parameters for first order, first-order plus time-delay (FOPTD) and second-order plus time-delay (SOPTD) model are given in **Table 3**.

[Supporting material of process identification is given in a separate compressed file (excel, python, Aspen Plus/Dynamics and MATLAB-Simulink programs) and readers can access files from the authors home page (https://sites.google.com/site/bcs12614/)].

From data fit and θ/τ values in **Table 3**, it can be inferred that g_{11} is best fit by the FO model whereas g_{12}, g_{21}, and g_{22} are best fit by the SOPTD model. The non-linear model, under the constraint given subsequently, can be represented by the transfer function given by Eq. (7). This can also be referred to as the original plant transfer function model. For the non-linear model, the manipulated variable is varied within the given range and the corresponding molar purities are obtained in the given range.

$$Gp(s) = \begin{bmatrix} \dfrac{0.00449}{1.00226s + 1} & \dfrac{0.00040\,e^{-0.21647s}}{1.0116s^2 + 1.8510s + 1} \\[2em] \dfrac{0.00205\,e^{-0.65557s}}{1.065s^2 + 1.9988s + 1} & \dfrac{0.00052\,e^{-0.40216s}}{4.1925s^2 + 3.07522s + 1} \end{bmatrix} \tag{7}$$

Model		First order		FOPTD		SOPTD	
		(GRG – Excel)		(Opt – Python)		(Opt – Python)	
g_{11}	K_p	0.00449		0.0045		0.0045	
	τ_p	1.00226		1.43951		0.67663	
	θ	—		0.33062		0.21039	
	ξ	—		—		1.10372	
	SSE	0.01054		0.005402		0.005268	
g_{12}	K_p	0.0004		0.0004		0.0004	
	τ_p	1.12152		1.00226		1.00579	
	θ (s)	—		0.01		0.21647	
	ξ	—		—		0.92018	
	SSE	0.00127		0.000296		0.000288	
g_{21}	K_p	0.00203		0.00205		0.00205	
	τ_p	1.00058		2.26224		1.03203	
	θ	—		0.69118		0.65557	
	ξ	—		—		0.96838	
	SSE	0.004382		0.004048		0.00407	
g_{22}	K_p	0.00053		0.00053		0.00052	
	τ_p	2.83704		2.67442		2.04758	
	θ	—		0.52164		0.40216	
	ξ	—		—		0.75094	
	SSE	≈ 0		≈ 0		≈ 0	
θ/τ		—		$\begin{bmatrix} 0.2297 & 0.01 \\ 0.3055 & 0.1950 \end{bmatrix}$		$\begin{bmatrix} 0.3109 & 0.2152 \\ 0.6352 & 0.1964 \end{bmatrix}$	
RGA		$\begin{bmatrix} 1.5180 & -0.5180 \\ -0.5180 & 1.5180 \end{bmatrix}$		$\begin{bmatrix} 1.5240 & -0.5240 \\ -0.5240 & 1.5240 \end{bmatrix}$		$\begin{bmatrix} 1.5240 & -0.5240 \\ -0.5240 & 1.5240 \end{bmatrix}$	
NI		0.6588		0.6562		0.6562	

Table 3.
Transfer function model parameters.

$$\left(\begin{array}{c} 886.237\ kW < u_1 < 899.631\ kW \\ -293.240\ kW < u_2 < -299.164\ kW \end{array} \right)$$

$$\left(\begin{array}{c} 0.9594 < y_1 < 0.9999 \\ 0.9926 < y_2 < 0.9993 \end{array} \right)$$

To evaluate the open-loop dynamic interactions between the PVs and MVs, the relative gain array (RGA) and the Niederlinski Index (NI) are applied [19]. The RGA for the RD-DPC process is:

$$RGA(\wedge) = \begin{bmatrix} 1.5413 & -0.5413 \\ -0.5413 & 1.5413 \end{bmatrix} \tag{8}$$

The NI for the original linearized plant model is 0.6488.

5.3 Closed-loop analysis

5.3.1 Controller settings

The decentralized controller settings are calculated by using IMC tuning relations [20, 21], and the centralized controller settings are calculated using TL tuning relations. The controller transfer functions are given in **Table 4**.

5.3.2 Simulations on linear and real non-linear model

The simulations on the linear model are carried out in MATLAB SIMULINK, first by employing the decentralized controller and then the centralized controller.

Decentralized controller		$G_{c\text{-}D}$
FO	PI	$\begin{bmatrix} 12.3732\left(1+\dfrac{1}{1.0023s}\right) & - \\ - & 104.8218\left(1+\dfrac{1}{2.8370s}\right) \end{bmatrix}$
FOPTD	PI	$\begin{bmatrix} 17.7403\left(1+\dfrac{1}{1.4395s}\right) & - \\ - & 155.3488\left(1+\dfrac{1}{2.6744s}\right) \end{bmatrix}$
	PID	$\begin{bmatrix} 17.7403\left(1+\dfrac{1}{1.4395s}\right) & - \\ - & 155.3488\left(1+\dfrac{1}{2.6744s}\right) \end{bmatrix}$
SOPTD	PID	$\begin{bmatrix} 59.9937\left(1+\dfrac{1}{1.4936s}+0.3065s\right) & - \\ - & 391.6324\left(1+\dfrac{1}{3.0752s}+1.3633s\right) \end{bmatrix}$
Centralized controller		$G_{c\text{-}C}$
FO	PI	$\begin{bmatrix} 18.7819\left(1+\dfrac{1}{1.3165s}\right) & -14.1750\left(1+\dfrac{1}{3.3303s}\right) \\ -71.9384\left(1+\dfrac{1}{3.7328s}\right) & 159.1149\left(1+\dfrac{1}{3.7266s}\right) \end{bmatrix}$
FOPTD	PI	$\begin{bmatrix} 17.7404\left(1+\dfrac{1}{1.4395s}\right) & -0.0110\left(1+\dfrac{1}{3.8412s}\right) \\ -1.7261\left(1+\dfrac{1}{1.7018s}\right) & 155.3498\left(1+\dfrac{1}{2.6744s}\right) \end{bmatrix}$
	PID	$\begin{bmatrix} 22.1556\left(1+\dfrac{1}{1.6048s}+0.1483s\right) & -0.0084\left(1+\dfrac{1}{4.6766s}+7.0832s\right) \\ -1.1484\left(1+\dfrac{1}{1.8063s}+0.1175s\right) & 190.8056\left(1+\dfrac{1}{2.9352s}+0.2377s\right) \end{bmatrix}$
SOPTD	PID	$\begin{bmatrix} 104.1324\left(1+\dfrac{1}{1.7074s}+0.4503s\right) & -66.5370\left(1+\dfrac{1}{2.8366s}+1.1245s\right) \\ -450.9374\left(1+\dfrac{1}{2.6268s}+1.1533s\right) & 679.7656\left(1+\dfrac{1}{3.5153s}+2.0050s\right) \end{bmatrix}$

Table 4.
Controller transfer functions.

Here, setpoint tracking (servo problem) and load rejection (regulator problem) simulations to the linearized plant transfer function model are carried out. The setpoint tracking is done by giving setpoint changes in y_{r1} and y_{r2}. y_{r1} is the setpoint to the controlled variable y_1 and similarly y_{r2} is the setpoint to y_2. Similarly, the disturbances are set to the input variable (u_1 and u_2) of the process.

The simulations on the non-linear model are done by replacing the linearized plant transfer function model with the original non-linear model. Here, the same controller settings of the linear model are applied along with base value (i.e., $u_{i,0} + \Delta u_i$) to the non-linear model in order to check the controller performance. The setpoint tracking is carried out by changing the setpoint in the range of 0.921 to 0.996 for y_{r1}, whereas y_{r2} is changed between 0.995 and 0.999. Particularly for the present case, the setpoints were set at 0.975 for y_{r1} and 0.996 for y_{r2}. Similarly, disturbances for the real model were considered as the feed flow rate of PA to the RD column (d_1) and reboiler heat duty of the separation column (d_2). The disturbances are in the range of $10\ kmol/hr < d_1 < 10.0185\ kmol/hr$, and $296.697\ kW < d_2 < 297.697\ kW$.

Linear model: Figures 4 and **5** show the response for all the transfer function models to setpoint changes and load changes, respectively, indicating the SOPTD model-based controller giving the best load rejection, less settling time, and reduced interactions. Similarly, **Figure 6** shows the comparative performance of the decentralized and the centralized SOPTD-PID controller for setpoint change, indicating centralized controller giving best performance.

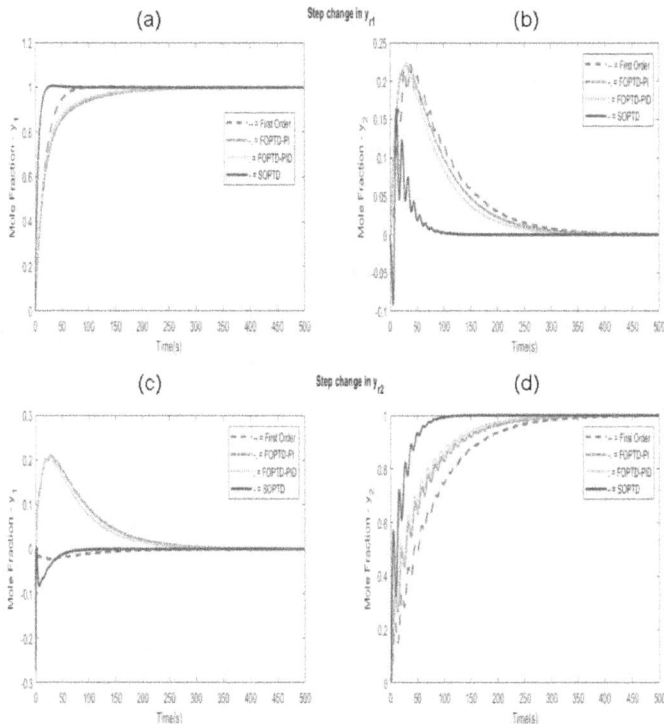

Figure 4.
Centralized controller – setpoint tracking and interactions for change in y_{r1} and y_{r2}. (a) and (d) represent the responses in y_1 and y_2, respectively, for the setpoint change in y_{r1} and y_{r2}. (b) and (c) represent the corresponding interactions.

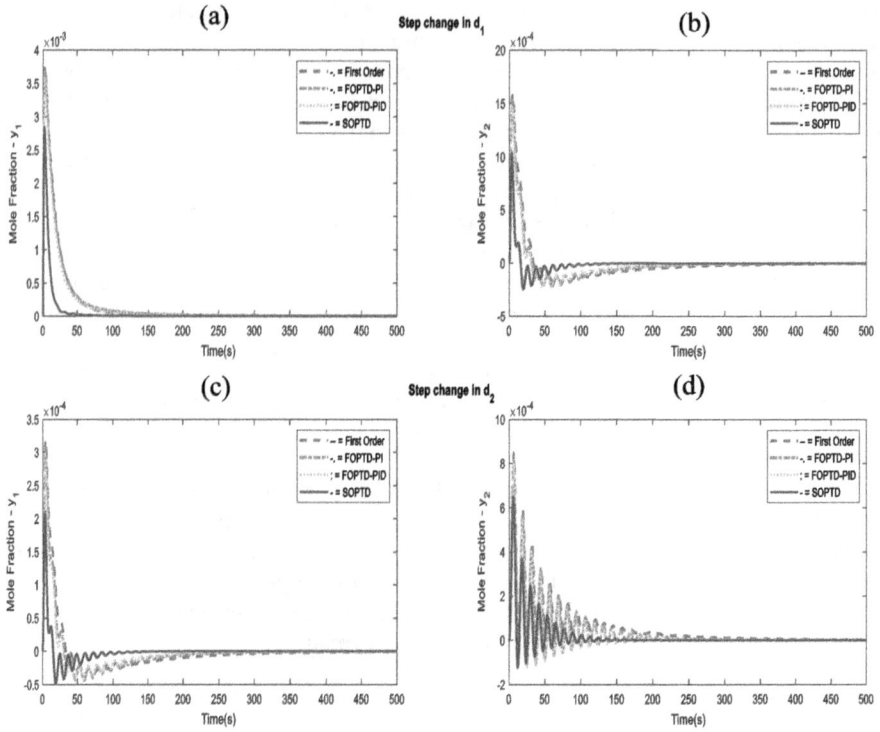

Figure 5.
Decentralized controller – load rejections and interactions for change in d_1 and d_2. (a) and (d) represent the responses in y_1 and y_2, respectively, for the load change in d_1 and d_2. (b) and (c) represent the corresponding interactions.

Non-linear model: Similarly, for non-linear model, it can be observed from **Figures 7–9** that SOPTD-PID centralized controller gives better load rejections and reduced interactions, as compared to other model-based controllers.

It is clear that both centralized and decentralized SOPTD-PID controllers show faster settling time, reduced interactions, and lower oscillations. From **Figures 6** and **9**, it is clear that the centralized controller gives faster settling, reduced interactions, and lower oscillations, as compared to the decentralized controller.

5.3.3 Robust stability analysis

The presence of model uncertainties necessitates the stability robustness of the multi-loop control system [22–26]. The dynamic perturbations existing in the system can be lumped into one single perturbation block Δ. To evaluate the robustness of the control system, inverse maximum singular value method is considered [17]. First, for a process multiplicative input uncertainty, $G(s)[I + \Delta_I(s)]$, the closed-loop system is stable if:

$$\|\Delta_I(j\omega)\| <_- \frac{1}{\bar{\sigma}} \left\{ [I + G_D(j\omega)G(j\omega)]^{-1} G_C(j\omega)G(j\omega) \right\} \qquad (9)$$

where $\bar{\sigma}$ is the maximum singular value of the closed-loop system. Similarly, for process multiplicative output uncertainty, $[I + \Delta_O(s)]G(s)$, the closed-loop system is stable if:

(a) SOPTD - Step change in y_{r1} (b)

(c) SOPTD - Step change in y_{r2} (d)

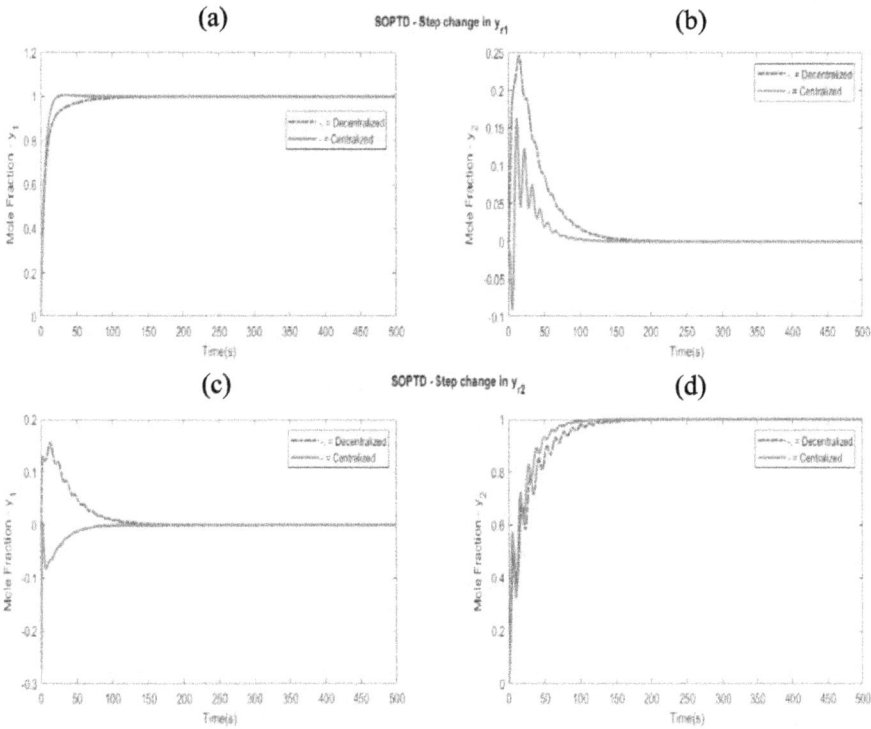

Figure 6.
SOPTD-PID controller – setpoint tracking and interactions for a given step change in y_{r1} and y_{r2}. (a) and (d) represent the responses in y_1 and y_2, respectively, for the setpoint change in y_{r1} and y_{r2}. (b) and (c) represent the corresponding interactions.

$$\|\Delta_O(j\omega)\| < \frac{1}{\sigma}\left\{[I + G(j\omega)G_C(j\omega)]^{-1}G(j\omega)G_C(j\omega)\right\} \tag{10}$$

The closed-loop system stability bounds are indicated by the frequency plots for the right-hand side part of Eqs. (9) and (10). The controller stability can be easily compared by comparing the area under the curve (more the area, more is the stability).

Figures 10 and **11** show the stability bounds for decentralized and centralized RD-DPC control, respectively. In these figures, the region above the curve indicates the instability region and that below the curve indicates the stable region. From **Figures 10** and **11**, it is clear that the FO-PI controller has more area under the curve, as compared to other controllers. Thus, the FO-PI controller gives robust control as compared to others, but this contradicts the above conclusions of SOPTD-PID controller performance being the best model. This can be explained as follows: For any magnitude of change to setpoint and lower magnitudes for disturbances, the SOPTD model-based controller gives the best performance. However, if the magnitude of disturbances is high, the first-order model-based controller gives the best performance. This can be easily interpreted from **Figures 12** and **13**. **Figure 12** shows that for lower frequency range (10^{-2} to 1 *rad/s*), the centralized FO-PI controller gives better robust stability as compared to the decentralized FO-PI controller. Similarly, from **Figure 13**, SOPTD-PID decentralized controller shows better robustness for higher frequency range (1 *rad/s* and above).

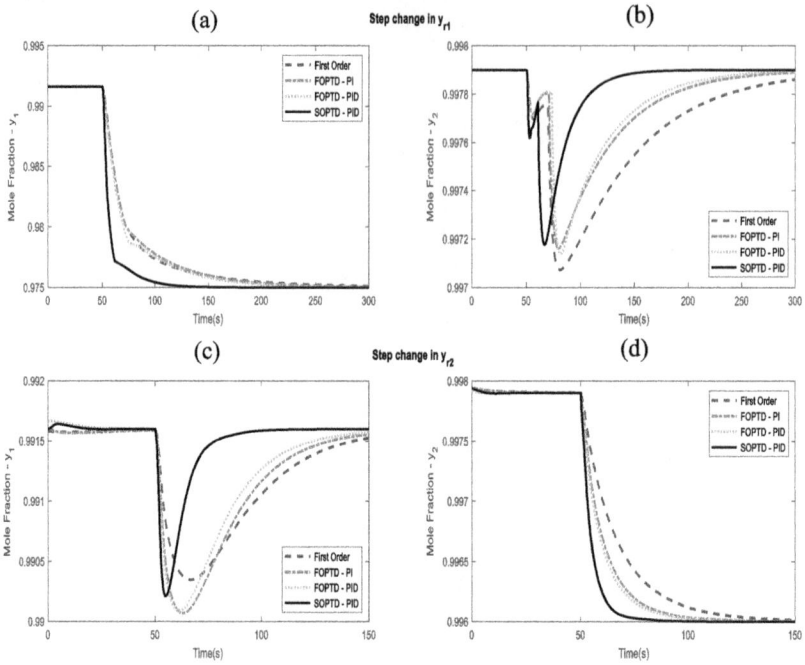

Figure 7.
Decentralized controller – setpoint tracking and interactions for a given step change in y_{r1} and y_{r2}. (a) and (d) represent the responses in y_1 and y_2, respectively, for the setpoint change in y_{r1} and y_{r2}. (b) and (c) represent the corresponding interactions.

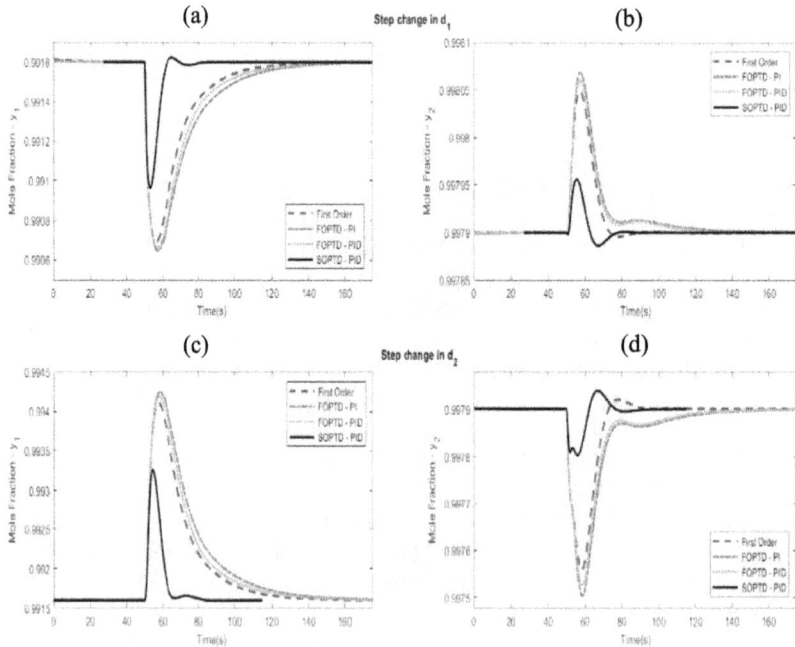

Figure 8.
Centralized controller – load rejections and interactions for a given step change in d_1 and d_2. (a) and (d) represent the responses in y_1 and y_2, respectively, for the load change in d_1 and d_2. (b) and (c) represent the corresponding interactions.

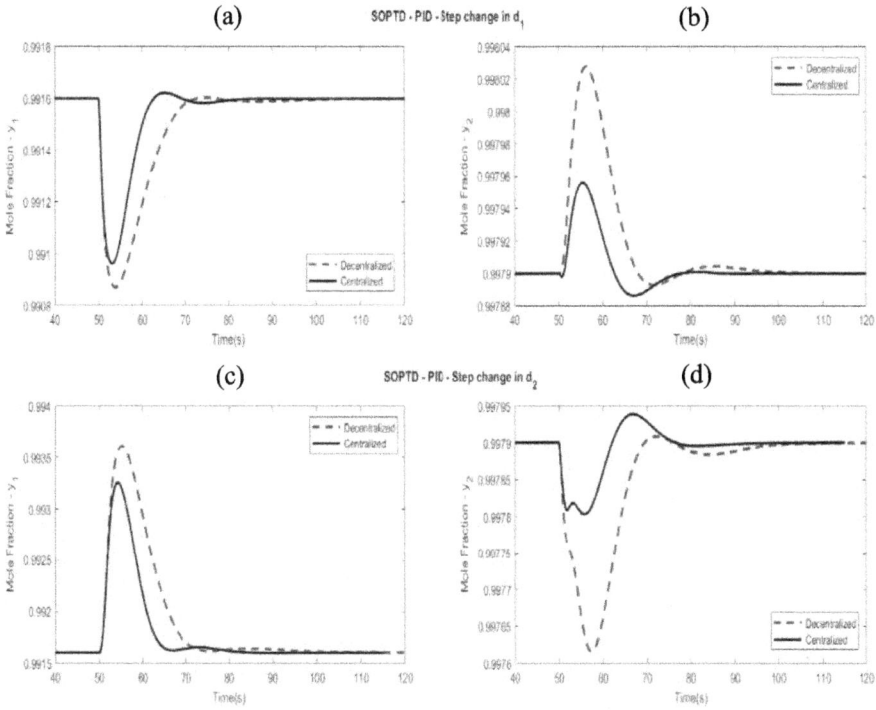

Figure 9.
SOPTD-PID controller – load rejections and interactions for a given step change in d_1 and d_2. (a) and (d) represent the responses in y_1 and y_2, respectively, for the load change in d_1 and d_2. (b) and (c) represent the corresponding interactions.

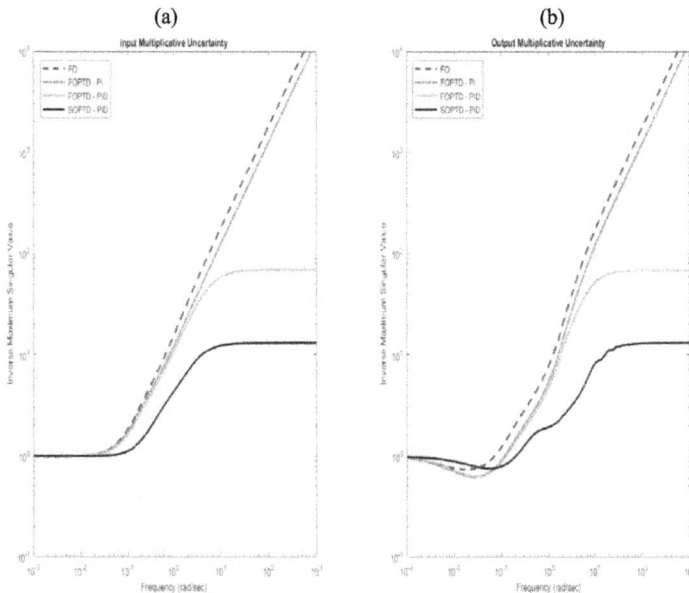

Figure 10.
Decentralized controller—Robustness—(a) input multiplicative and (b) output multiplicative uncertainties.

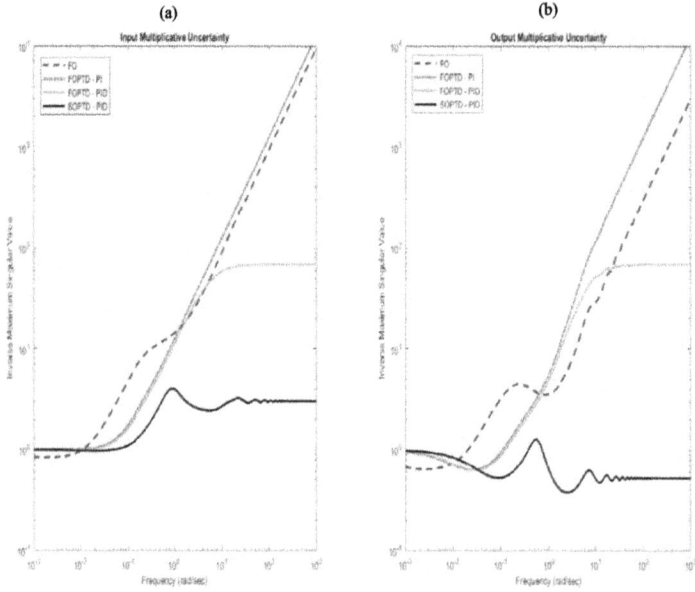

Figure 11.
Centralized controller—Robustness—(a) input multiplicative and (b) output multiplicative uncertainties.

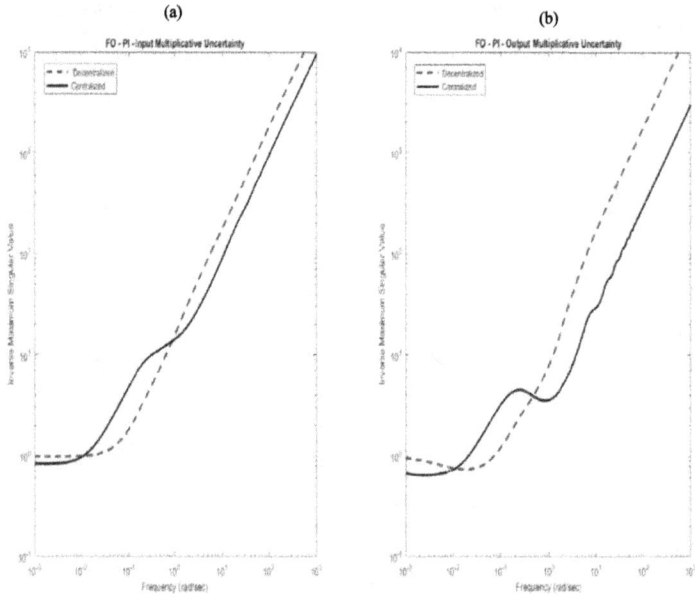

Figure 12.
FO-PI controller—Robustness—(a) input multiplicative and (b) output multiplicative uncertainties.

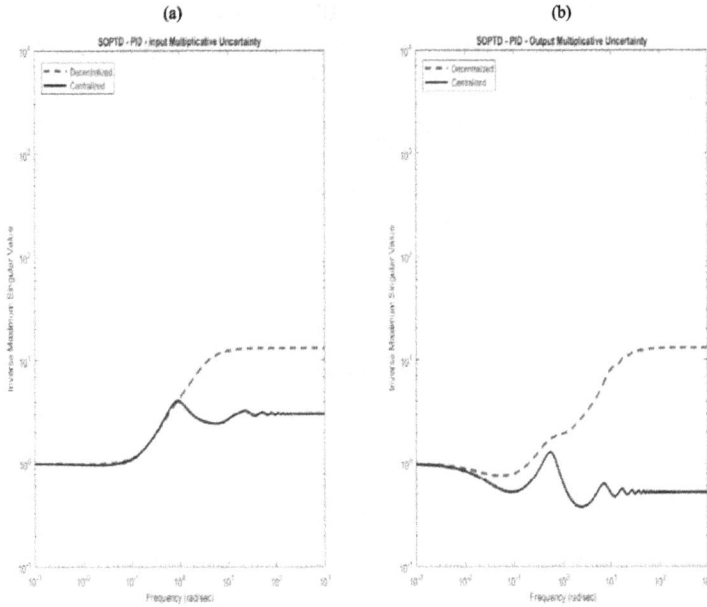

Figure 13.
SOPTD-PID controller—Robustness—(a) input multiplicative and (b) output multiplicative uncertainties.

6. Conclusion

This chapter discusses the presence of multiple steady states indicating non-linear reactive distillation process. The presence of multiple steady states urged to fix the operating range for the manipulated variable. For achieving the purity of DPC (y_1), more than 99% of the reboiler heat duty of the RD column (u_1) must be varied between 886.237 kW and 899.631 kW. Similarly, for purity of MA (y_2), more than 99% of the condenser heat duty of the separation column (u_2) must be constrained between −293.240 kW and − 299.164 kW. This smaller operating range shows that the process is highly sensitive. Furthermore, if we start up the plant avoiding the given operating ranges, we end up at another steady state and hence undesirable product purities. The controller settings, derived from IMC and TL tuning relations, when applied to linear model as well as to non-linear model show proper setpoint tracking and load rejections. From the quantitative performance measures, setpoint tracking, and load rejection tests, the SOPTD–PID controller gives the best performance amongst the FO-PI, FOPTD-PI, FOPTD-PID, and SOPTD-PID controllers. Also, the centralized controller gives better performance, as compared to decentralized controller. Thus, the centralized controller regulates away the interactions more effectively than the decentralized controller. Even for $\lambda ij > 1$, the centralized controller shows better performances. However, the SOPTD model-based controller gives the best performance for any magnitude of setpoint value change and low value of load change. If the load value is high, FO model-based controller gave the best performance, as indicated by robust stability analysis.

In other words, for lower frequency range (10^{-2} to 1 rad/s), the centralized FO-PI controller gives better robust stability as compared to decentralized FO-PI controller, and SOPTD-PID decentralized controller shows better robustness for higher frequency range (1 rad/s and above). We also conclude that the setpoint changes are tracked effectively for higher order models, whereas the load changes may or may not be regulated by higher order models. Thus, a proper trade-off has to be done between performance and robustness when selecting the control configuration and the model-based controller.

Conflict of interest

The authors declare no conflict of interest.

Author details

Shirish Prakash Bandsode and Chandra Shekar Besta*
Department of Chemical Engineering, National Institute of Technology, Calicut, Kerala, India

*Address all correspondence to: schandra@nitc.ac.in

IntechOpen

References

[1] Cheng K, Wang SJ, Wong DSH. Steady-state design of thermally coupled reactive distillation process for the synthesis of diphenyl carbonate. Computers and Chemical Engineering. 2013;**52**(June):262-271

[2] Haubrock J. The process of dimethyl carbonate to diphenyl carbonate: Thermodynamics, reaction kinetics and conceptional process design [thesis]. Enschede: University of Twente; 2007

[3] Hung S-B, Tang Y-T, Chen Y-W, Lai I-K, Hung W-J, Huang H-P, et al. Dynamics and control of reactive distillation configurations for acetic acid esterification. IFAC Proceedings Volumes. 2006;**39**(2):403-408. DOI: 10.3182/20060402-4-BR-2902.00403

[4] Pavan Kumar MV, Kaistha N. Steady-state multiplicity and its implications on the control of an ideal reactive distillation column. Industrial & Engineering Chemistry Research. 2008;**47**(8):2778-2787

[5] Nandong J, Samyudia Y, Tadé MO. Control structure analysis and design for nonlinear multivariable systems. IFAC Proceedings Volumes. 2007;**40**(5):251-256

[6] Lengare MJ, Chile RH, Waghmare LM. Design of decentralized controllers for MIMO processes. Computers & Electrical Engineering [Internet]. 2012;**38**(1):140-147. DOI: 10.1016/j.compeleceng.2011.11.027

[7] Jaibhavani KS, Hannuja B. Modeling and development of decentralized PI controller for TITO system. In: Proceedings - TIMA 2017; 9th International Conference on Trends in Industrial Measurement and Automation; 6-8 January 2017; Chennai. New Jersey: IEEE; 2017. pp. 45-48

[8] Tavakoli S, Griffin I, Fleming PJ. Tuning of decentralised PI (PID) controllers for TITO processes. Control Engineering Practice. 2006;**14**(9):1069-1080

[9] Ravi VR, Thyagarajan T. A decentralized PID controller for interacting non linear systems. In: 2011 International Conference on Emerging Trends in Electrical and Computer Technology; 23-24 March 2011; Nagercoil, India. IEEE; 2011. pp. 297-302

[10] Besta CS, Chidambaram M. Decentralized PID controllers by synthesis method for multivariable unstable systems. IFAC-PapersOnLine [Internet]. 2016;**49**(1):504-509. DOI: 10.1016/j.ifacol.2016.03.104

[11] Shen Y, Sun Y, Xu W. Centralized PI/PID controller design for multivariable processes. Industrial & Engineering Chemistry Research. 2014;**53**(25):10439-10447

[12] Park BE, Sung SW, Lee IB. Design of centralized PID controllers for TITO processes. In: 2017 6th International Symposium on Advanced Control of Industrial Process AdCONIP; 28-31 May 2017; Taipei. IEEE; 2017. pp. 523–528

[13] Swetha M, Kiranmayi R, Swathi N. Design of centralized PI control system for two variable processes based on root locus technique. International Journal of Innovative Technology and Exploring Engineering. 2019;**9**(1):4851-4855

[14] Besta CS, Chidambaram M. Design of centralized PI controllers by synthesis method for TITO systems. Indian Chemical Engineer. 2017;**59**(4):259-279

[15] Chen Q, Luan X, Liu F. Analytical design of centralized PI controller for high dimensional multivariable systems [Internet]. IFAC Proceedings Volumes.

2013;**46**(32):643-648. DOI: 10.3182/
20131218-3-IN-2045.00030

[16] Skogestad S. Simple analytical rules
for model reduction and PID controller
tuning. Modeling, Identification and
Control. 2004;**25**(2):85-120

[17] Besta CS, Chidambaram M. Tuning
of multivariable PI controllers by BLT
method for TITO systems. Chemical
Engineering Communications
[Internet]. 2016;**203**(4):527-538.
DOI: 10.1080/00986445.2015.1039121

[18] Besta CS, Chidambaram M.
Modelling of interactive multivariable
systems for control. In: Regupathi I,
Shetty V, Thanabalan M, editors. Recent
Advances in Chemical Engineering.
Singapore: Springer; 2009. pp. 285-291

[19] Chidambaram M, Saxena N. Relay
Tuning of PID Controllers [Internet].
Singapore: Springer; 2018.
DOI: 10.1007/978-981-10-7727-2

[20] Besta CS. Control of unstable
multivariable systems by IMC method.
In: 2017 Trends in Industrial
Measurement and Automation (TIMA);
6-8 January 2017; Chennai. IEEE; 2017.
pp. 0–5

[21] Besta CS, Chidambaram M.
Improved decentralized controllers for
stable systems by IMC method. Indian
Chemical Engineer [Internet]. 2018;
60(4):418-437. DOI: 10.1080/
00194506.2017.1280422

[22] Dileep D, Michiels W, Hetel L,
Richard JP. Design of robust structurally
constrained controllers for MIMO plants
with time-delays. In: 2018 European
Control Conference ECC; 12-15 June
2018; Limassol, Cyprus. IEEE; 2018.
pp. 1566-1571

[23] Panyam Vuppu GKR, Makam
Venkata S, Kodati S. Robust design of
PID controller using IMC technique for
integrating process based on maximum

sensitivity. Journal of Control,
Automation and Electrical Systems.
2015;**26**(5):466-475

[24] Zhu ZX, Jutan A. Consistency
principles for stability in decentralized
control systems. Chemical Engineering
Communications. 1995;**132**(1):107-123

[25] Dittmar R, Gill S, Singh H, Darby M.
Robust optimization-based multi-loop
PID controller tuning: A new tool and its
industrial application. Control
Engineering Practice [Internet]. 2012;
20(4):355-370. DOI: 10.1016/j.
conengprac.2011.10.011

[26] Taiwo O, Adeyemo S, Bamimore A,
King R. Centralized robust multivariable
controller design using optimization.
IFAC Proceedings Volumes. 2014;**19**:
5746-5751